POLONIUM IN THE PLAYHOUSE

POLONIUM
IN THE PLAYHOUSE

*The Manhattan Project's Secret
Chemistry Work in Dayton, Ohio*

LINDA CARRICK THOMAS

Trillium, an imprint of
The Ohio State University Press
Columbus

Library of Congress Cataloging-in-Publication Data
Names: Thomas, Linda Carrick, 1960– author.
Title: Polonium in the Playhouse : the Manhattan Project's secret chemistry work in
    Dayton, Ohio / Linda Carrick Thomas.
Description: Columbus : Trillium, an imprint of The Ohio State University Press,
    [2017] | Includes bibliographical references and index.
Identifiers: LCCN 2016059022 | ISBN 9780814213384 (cloth ; alk. paper) |
    ISBN 0814213383 (cloth ; alk. paper)
Subjects: LCSH: Manhattan Project (U.S.)—History. | Thomas, Charles Allen, 1900–
    1982. | Polonium—Research—Ohio—Dayton—History.
Classification: LCC QC773.3.U5 T46 2017 | DDC 355.8/25119097309044—dc23
LC record available at https://lccn.loc.gov/2016059022

Cover design by Angela Moody
Text design by Juliet Williams
Type set in Palatino and Myriad

Cover photos: National Archives (Atlanta), records of the Atomic Energy Commission,
and the Talbott family

9 8 7 6 5 4 3 2 1

Dedicated to Lyle F. Albright (1921–2010)

Professor Emeritus of Chemical Engineering, Purdue University
Manhattan Project veteran—Hanford Engineer Works, health physics

*For his encouragement and contagious determination that the public
understand the vast spread of the Manhattan Project and recognize the
thousands of men and women who labored at its sites across the country.*

The battle of the laboratories held fateful risks for us as well as the battles of the air, land, and sea, and we have now won the battle of the laboratories as we have won the other battles.

—President Harry Truman, August 6, 1945

It is one of the ironies of history that an episode only begins to seem important at a time when memories have begun to fade, and when some of those who participated can no longer tell their stories.

—Alice Kimball Smith, *Bulletin of the Atomic Scientists,* 1958

CONTENTS

APPENDICES

ILLUSTRATIONS

PREFACE

THE MANHATTAN PROJECT was a vast $2 billion undertaking that involved more than 125,000 workers at its peak in 1944, with some 69,000 people involved in operations and research.[1] The project to build an atomic bomb was far more than the sites in Los Alamos, Oak Ridge, and Hanford. Scientists, engineers, and support staff worked at more than 30 locations across the country, from industrial plants to academic laboratories to government facilities.[2] All told, hundreds of entities contributed to the project.[3] The work in Dayton was just one piece of the puzzle.

The work at all sites was focused on research and development of two types of atomic bombs. The uranium bomb, nicknamed "Little Boy," was dropped on Hiroshima, Japan, on August 6, 1945. Two plutonium bombs were made. The first, known as the "Gadget," was tested at the Trinity Site in New Mexico on July 16, 1945. The second, "Fat Man," was dropped on Nagasaki on August 9, 1945. A third was prepared and ready for use, but Japan surrendered on August 15, 1945.

The chemistry of the Manhattan Project focused on studying the materials to be used in the bombs—learning about the newly discovered plutonium—and developing processes for refining them. It was an enormous, complicated undertaking that led to a remarkable number of discoveries and scientific advances in a short amount of time. With some 2,500 people on staff at the peak of the Project, Los Alamos employed 297 workers in its chemistry and metallurgy (CM) division in March of 1945; six of the scientists focused on polonium.[4] The Los Alamos CM staff was also responsible for work with uranium, plutonium, carbides and tampers, monitoring and decontamination, and general service chemistry for other divisions.[5] The University of Chicago's Metallurgical Laboratory (Met Lab), with 2,000 staff members at its peak in July 1944, had 99 assigned to C-I, the chemistry division responsible for plutonium purification.[6] Dayton operations, which focused solely on polonium, had a staff of 201 in 1945. Manhattan Project chemists and metallurgists pursued multiple lines of inquiry and explored many different processes as they developed techniques to produce the materials for the atomic bombs. While this book discusses in basic terms some of the chemistry involved in the project, readers will find more detailed scientific information in other sources; some still remain classified.[7]

In May of 1943, as the Manhattan Project was organized, attention turned to the chemistry challenges of the atomic bomb effort. Scientific director J. Robert Oppenheimer had envisioned a small chemistry section in Los Alamos that would provide service to the physicists. The bulk of the theoretical chemistry work would be undertaken at the University of Chicago and the University of California, Berkeley. That spring, however, it became apparent that the chemistry would need to be more centrally administered and that the number of chemists on the project would have to increase. The scope of the bomb project had grown from a laboratory-sized endeavor to one of near-industrial scale. Facilities capable of producing mass amounts of plutonium and polonium would require industrial processes and management skills. DuPont had already been enlisted to construct and manage the reactor facilities in Tennessee and Washington State, but an administrator familiar with both science and industry was needed to oversee the chemistry and metallurgy of the project. Industrial chemist Charles Allen Thomas, director of Monsanto Chemical Company's central research facility in Dayton, Ohio, was the man who fit the bill. Thomas was my paternal grandfather; my father, Charles Allen Thomas III, was the oldest of his four children.

ONE MAN, TWO JOBS

In May 1943, Thomas was appointed coordinator of Manhattan Project chemistry and metallurgy. In this capacity, he managed multiple key assignments during his tenure with the Project from 1943 to 1945.

P.1. Manhattan Project chemistry sites. DOE, adapted by author.

The first task was coordination of plutonium chemistry and met-allurgy at various Manhattan Project sites: the University of Chicago's Metallurgical Laboratory (Met Lab); the University of California, Berkeley's Radiation Laboratory (Rad Lab); Los Alamos (Site Y); and Iowa State College in Ames, where work focused on uranium. The second task assigned to Thomas was part of the chemistry and metal-lurgy work—oversight of polonium purification at Monsanto Chemi-cal Company facilities in Dayton, Ohio.

In 2004, representatives of the National Park Service (NPS) in Day-ton contacted my father for support in promoting the inclusion of Dayton Project sites in a new Manhattan Project National Park then under consideration.[8] As a journalist by training and practice, and as a resident of nearby Indiana, I was appointed the Thomas family liai-son with the NPS. The greater family also saw this as a good time for me to pull the history of my grandfather's work together into a book for the family. After meeting with NPS officials, Dayton politicians, and Dayton Project veterans, it became clear to me that the history of the work in Dayton was of much greater importance than a book for the family. The story needed to be shared with the public.

I began reviewing existing histories and government documents on the Manhattan Project, consulted the Atomic Heritage Foun-dation's vast online resource of Manhattan Project oral histories,

reviewed my grandfather's personal papers, and reached out to historical societies, archivists, historians, and the very few living Dayton
Project veterans I could find. I made a Freedom of Information Act
request for records from Los Alamos National Laboratory and buried myself in all things Dayton Project. Although I uncovered ample
information for a book, I found that existing Manhattan Project literature barely—usually as a footnote or a sentence or two—mentioned the work in Dayton. I was puzzled about why one hadn't been
written. There are many possible explanations: (1) the chemistry of
the Manhattan Project—especially that undertaken in Dayton—has
remained one of the most highly classified portions of the bomb
project because it included new and carefully guarded discoveries
and processes. Nuclear physics was more widely known at the time;
(2) Dayton was considered an auxiliary site; and (3) the official government document reporting the history of the Dayton Project was
not declassified until July 13, 1983, with portions of the polonium
work remaining classified today. As for Thomas, who was nearly as
high ranked as Oppenheimer and the other Manhattan Project leaders
and interacted with them on the most classified levels, he somehow
stayed out of historical reports. Whatever the reasons, adequate information on the work of Thomas and the Dayton Project is missing in
most general histories of the Manhattan Project.

Over the short course of the Manhattan Project, many different approaches to achieving purified plutonium and polonium
were developed, attempted, and aborted. Purity levels were set and
changed. Quantities needed for the bombs were agreed upon and
then jettisoned. Delivery dates were established and then resched-
uled. Work in Dayton was as intensive and as constantly changing
as that at other sites across the Manhattan Project, as General Groves,
Oppenheimer, and their team of advisors considered the work of the
theoretical scientists and sought the fastest and best way to build
an atomic bomb. Accordingly, the story of the Dayton Project and
the man chosen to coordinate the essential work cannot be told in a
smooth, chronologically continuous manner. As a historical report of
the Dayton Project points out, "an attempt to write . . . in such a way
as to give a true chronological picture of the various steps involved in
the development of production methods would only result in mental
chaos to the uninitiated reader."[9]

This book will lay the foundation for the polonium research and development work in Dayton by setting the stage in an industrial chemical laboratory. Written for general audiences, it does not focus on technical detail. A primer on the basic science can be found in Appendix I. The book begins by introducing Thomas, who shared many characteristics with other Manhattan Project leaders—a relentless curiosity, a drive to succeed against all odds, and an all-American success story in the making. Then it will introduce his pre-war work as the co-founder and president of the nation's largest independent consulting laboratory and its purchase by Monsanto. And with that foundation, it will tell how the man and the science came together to support the Manhattan Project. The story will wrap up with a look at the role that Dayton and Thomas played in atomic energy and international policy following the war.

My grandfather died in 1982 when I was 21. I was well aware of his accomplishments in industry, but not familiar with his Manhattan Project work; he never spoke about it. "As a result of the many years and layers of secrecy, no previous histories of Dayton or of the Manhattan Project have adequately synthesized the story of the Dayton Project," wrote National Park Service historian Ed Roach. "While the consequences of this research in Dayton impacted the entire world, the tight, enduring secrecy associated with the Project has denied Thomas and his colleagues recognition as contributors to one of the most complex and important achievements of the Manhattan Project."[10] This book will at long last cast light on these achievements.

ACKNOWLEDGMENTS

MUCH LIKE the Manhattan Project itself, this book was a communal effort, from the moment of inspiration to the finished product. In the decade spent on this project, I discovered and became part of a small but dedicated group of people whose lives in one way or another have been influenced by the Manhattan Project. Each of them helped me tell the story of the work in Dayton, Ohio. And to each of them, I am eternally grateful for the selfless, enthusiastic, and generous hand that was offered.

The work would not have been started without the interest of Dayton Aviation Heritage National Historical Park personnel, especially Timothy S. Good and Ed Roach, who began the work of gathering documents related to the Dayton Project. And it would not have been finished without the enthusiastic support of the Ohio State University Press—director Anthony Sanfilippo; managing editor Tara Cyphers; production assistant Debra Jul; and marketing director Laurie Avery.

Several Manhattan Project authors and historians supported this project. Robert S. Norris offered immeasurable help and encouragement as I worked to get the project from desktop to publisher. He was the first of several reviewers whose comments guided my manuscript toward its final phase. I am also indebted to Cameron Reed, Frank Settle, John Coster-Mullen, and Carey Sublette for the time and care each took reviewing my text or providing technical images.

Lee Curtis lent a skillful eye as a copy editor and research assistant. Jessie Farrington offered keen insight on content flow. And Greg Schultz exhibited great skill at locating images and tracking permissions, which saved me too many times to count. Photographer Mark Simons and designers Pamela Anderson and Dawn Minns reminded me that having talented and generous friends is such a gift.

Archivists Sonya Rooney at Washington University Libraries in St. Louis and Shane Bell at the National Archives in Atlanta went beyond the scope of their job descriptions to help me locate documents. Dayton-area archivists, including Mark W. Risley and David A. Schmidt of the Oakwood Historical Society, and Ray Seiler and Dick Madding of the Mound Science and Energy Museum, provided help with materials and questions about life in Dayton. Archivist Alan Carr of Los Alamos National Laboratory was always quick to respond to a call for help. And the Atomic Heritage Foundation proved to be an invaluable resource for oral histories of Manhattan Project veterans. Thank you to each of them.

It was an immense pleasure and honor to work with Dayton Project veterans George Mahfouz, who hosted me on a visit to Dayton in 2005, and Betty (Halley) Jones. The families of veterans also rallied to my call, including Emma De Benedetti, wife, and Vera Bonnet, daughter, of Sergio De Benedetti; Elizabeth Sopka, daughter of John Sopka; Robert Weimer, son of Harry Weimer; and Bill Curtis, son of Mary Lou Curtis.

Last but in no way least is a debt of incalculable gratitude to my family. Talbott cousins Ed Schultz and Mimi Mead-Hagen dug deep into family archives. My children—Carrick, Henry, and Charlie—patiently tolerated my decade-long distraction with researching and writing this book, obliged me by serving as test audiences and reading through drafts of copy, and cheered me from the sidelines. My parents, Charles Allen Thomas III and Margaret Thomas, shared insight, memories, wisdom, opinions, and encouragement. And my husband, Greg, exhibited extreme patience, calm, support, and acceptance of the focus required to produce this book.

As more than one fellow Manhattan Project author told me, writing this type of book is a labor of love. And it is indeed with the loving support of all that this book has been written.

GLOSSARY OF TERMS AND NAMES

TERMS

49   Code for plutonium, element 94.

Clinton   Clinton Engineer Works, Tennessee (now known as Oak
                   Ridge). During the Manhattan Project, the site contained
                   two facilities that produced enriched uranium—the K-25
                   gaseous diffusion plant and the Y-12 electromagnetic
                   enrichment plant. A third facility, X-10, produced plu-
                   tonium. It was based on Fermi's experimental CP-1 and
                   informed the design of the production reactor at the Han-
                   ford site. The first batch of plutonium was refined in Clin-
                   ton's 221-T Plant.

CP-1   Chicago Pile-1, University of Chicago experimental
                   reactor.

Cups   Code name for curies.

Met Lab   University of Chicago Metallurgical Laboratory.

Polonium   A by-product of the lead-containing wastes from
                   uranium-, vanadium-, and radium-refining operations.
                   It can also be synthetically manufactured by processing
                   irradiated bismuth.

Postum   Manhattan Project code name for polonium, sometimes
                   misspelled as *potsum.*

Site W   Hanford Engineer Works. Located near Richland, Wash-
                   ington. The 586-square-mile site contained three nuclear
                   reactors (105-B, 105-D, and 105-F) where uranium slugs

were irradiated to produce plutonium, and three chemical
separation plants where the plutonium was separated.
Construction of the "B" reactor—the first large-scale plutonium production reactor in the world—began in October
1943 and was completed in September 1944. The reactor
was based on Fermi's experimental design and the pilot
plant at Clinton.

Site Y   The secret name for the Manhattan Project's central labora-
                       tory in what is now known as Los Alamos, New Mexico.

Unit I   Dayton Project headquarters, Nicholas Road.

Unit III   Dayton Project facility, 1601 W. First Street, Dayton. Also
                       known as Bonebrake after the building's origins as Bone-
                       brake Theological Seminary.

Unit IV   Polonium-processing facility at the intersection of 715
                       Runnymede Road and Dixon Road in Oakwood. Also
                       known as Runnymede Playhouse, or the Playhouse.

The Warehouse   A six-story warehouse at 601 E. Third Street in down-
                       town Dayton. Used as a health physics laboratory and for
                       storage.

NAMES

Arthur Holly Compton (AHC), physicist and Manhattan Project advisor and
    administrator, as well as project director of the University of Chicago's
    Metallurgical Laboratory (1942–45).

James B. Conant (JBC), chemist and president of Harvard University, chair-
    man of the National Defense Research Committee, advisor to General Les-
    lie R. Groves, member of Interim Committee (1945) that advised President
    Harry Truman on nuclear weapons.

Richard W. Dodson (RWD), chemist and Los Alamos radiochemistry group
    leader, then associate division chief for Los Alamos chemistry.

W. Conard Fernelius (WCF), chemist and Dayton Project assistant laboratory
    director, then associate laboratory director (July 1945) and laboratory
    director (November 1945).

Joseph W. Kennedy (JWK), chemist and co-discoverer of plutonium with
    Glenn Seaborg, Edwin McMillan, and Arthur Wahl; Division leader, Los
    Alamos Chemistry and Metallurgy.

General Leslie R. Groves (LRG), Army Corps of Engineers officer and director of the Manhattan Engineer District (Manhattan Project).

James H. Lum (JHL), Monsanto chemist and Dayton Project laboratory director, then executive director of Clinton Laboratories (1945–47).

J. Robert Oppenheimer (JRO), physicist and science director of the Manhattan Project.

Charles Allen Thomas (CAT), Monsanto Chemical Company vice president, coordinator of Manhattan Project chemistry and metallurgy, and director of the Dayton Project.

Charles Allen Thomas III (CAT III), oldest child of Charles Allen Thomas.

Frances Carrick Thomas (FCT), mother of Charles Allen Thomas.

CHAPTER 1

SETTING THE SCENE

DURING THE SPRING of 1944, delivery trucks began rolling in and out of the affluent Dayton neighborhood of Oakwood, navigating winding streets and passing leafy estates such as that of aviation pioneer Orville Wright. The trucks came and went at all hours from the Talbott Playhouse at 715 Runnymede Road, a prominent family's indoor tennis court and recreation facility. They drew the attention of neighbors who knew not to ask too much. It was wartime and the nation was involved in defense activity.

Residents were told the trucks were going to and from a government film processing facility. They were, in fact, couriers transporting one of the world's most dangerous elements, polonium, which was purified in Oakwood and then shipped to Los Alamos for use in the trigger of the plutonium bomb first tested at the Trinity site in the New Mexico desert and conclusively detonated over Nagasaki on August 9, 1945. The work, known as the Dayton Project, was the most closely guarded portion of the Manhattan Project because its product—refined polonium for the trigger—was so essential to the bomb. Its history is not widely known.

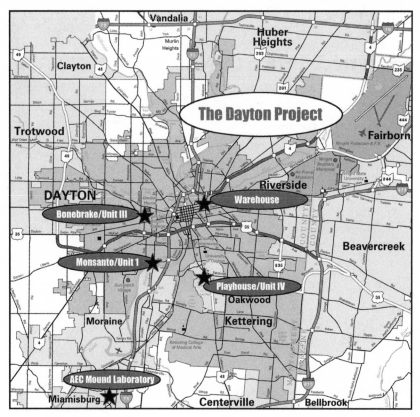

1.1. Dayton Project sites. Ohio Department of Transportation, adapted by Dawn M. Minns.

The work in Dayton was hidden in plain sight. The main Manhattan Project sites were built in remote areas—Site Y (Los Alamos) in the New Mexico desert, Clinton (now known as Oak Ridge) in the hills of Tennessee, and Hanford on the plains of Washington. The work in Dayton, by comparison, took place in the middle of a busy metropolitan area. Laboratories were located downtown in a former elementary school and on an estate in a residential neighborhood. Industrial chemist Charles Allen Thomas, the MIT-educated director of Dayton's Monsanto Central Research, directed this work. As coordinator of chemistry and metallurgy for the Manhattan Project, he also oversaw the plutonium research and development in laboratories at

1.2. Ohio map. Ohio Department of
Transportation.

Los Alamos; the University of California, Berkeley; the University of
Chicago's Metallurgical lab (Met Lab); and at Iowa State College.

The highly classified nature of the work in Dayton may well have
obscured its place in history. Shortly after the war, Manhattan Project director General Leslie Groves acknowledged this in a letter to
Monsanto chairman Edgar Queeny, stating that information on the
Dayton Project must remain publicly unreleased. "A detailed description of your efforts must still remain undisclosed because of security
considerations," Groves wrote.[1] The work was so essential and buried so deep within the overall project that Dayton Project personnel
could not obtain supplies through the usual priority rating given
Manhattan Project sites, lest the work be associated with the bomb
project. "It was considered very secret. It was part of the weapon
(but) we gave very little indication to anybody we needed it," recalled
Major General Kenneth Nichols, who served as Manhattan District
Engineer, overseeing the uranium and plutonium production at Oak
Ridge and Hanford.[2] Records for the Dayton Project remain among

the most highly classified of Manhattan Project documents and have not been readily available to those exploring the vast map of the Manhattan Project. It is as if the Dayton Project never existed. And yet it did.

DAYTON'S LEGACY OF INNOVATION

The Dayton Project's classified chemistry took place in relative obscurity in the middle of the United States. The city of Dayton lies north of Cincinnati in the Miami Valley of Southern Ohio. It is perhaps best known as the home of aviation pioneers Orville and Wilbur Wright, who brought flight to the world in 1903. In addition to the Wright Brothers, it was also home to an unusually high number of inventors and entrepreneurs for a city its size. In 1900, Dayton claimed more patents per capita than any other U.S. city.[3]

In the 1920s and '30s, Dayton was a bustling center of invention, with a population of just under 200,000 residents. Noteworthy residents included James Ritty (1836–1918), inventor of the mechanical cash register; John H. Patterson (1844–1922), founder of National Cash Register (NCR); engineer Charles F. "Boss" Kettering (1876–1958), father of the first electrical cash register, the electric starter for automobiles, and anti-knock gasoline, among many other inventions. Much of Kettering's work took place at Dayton Engineering Laboratories Co. (Delco), which was co-founded by Kettering and Edward Deeds, who had worked together at NCR. John Q. Sherman founded Standard Register in Dayton to produce the autographic register to duplicate documents. The city was also home to Frank M. Tait, considered the dean of America's electric industry. The world's first electric refrigerator, the Frigidaire, was built in Dayton, and Dayton Computing Scale merged with other units to become the nucleus of International Business Machines (IBM) Corporation.

The city also had a significant military presence. It was home to Wright Air Field (now Wright-Patterson Air Force Base) and the Army Air Service headquarters, the nation's first aviation research center. It was also home to the first Strategic Air Command, the precursor to the Air Force.[4] In 1917, a group of investors who anticipated the Army Air Service's demand in World War I for airplanes created the

Dayton-Wright Company. The company, co-owned by Kettering, Orville Wright, and businessman Harold E. Talbott, delivered 3,106 De Havilland DH-4 biplanes and 400 Standard SJ-1 trainers. By 1918, the consortium of companies owned by the team of investors (Delco, Dayton-Wright, and Dayton Metal Products Company) had become part of a larger effort, General Motors, with Kettering named vice president of General Motors Research Corporation in 1920.

In addition to Monsanto's Manhattan Project laboratories, other Dayton industries were also busy with wartime defense contracts. By 1943, some 61 principal war production industries in Dayton employed 115,000 people.[5] Dayton contributions included:[6]

- Serum refrigerators: Chrysler Corp. Airtemp division
- The "bombe," a machine used to break the German "E Code" (code sent and received by German U boats): NCR
- Precision gauges: Sheffield Corp.
- Hamilton variable pitch propeller, Hamilton propellors for B29s: General Motors Frigidaire division
- Synthetic rubber tires: Dayton Rubber Company
- Chandler-Evan carburetors for B24 Liberator bomber: NCR
- Bomber gears for B26 Martin Marauders: Delco
- .50 caliber machine gun/side gunner on Flying Fortress: Frigidaire
- M-30 carbine: General Motors Inland Division

With its legacy as a birthplace of ideas and inventions, a home to visionaries, an established military air base, a geographic location at the crossroads of the nation, and an understated international profile, Dayton was well suited for its role in the Manhattan Project.

CHAPTER 2

CHARLES ALLEN THOMAS— THE MAKING OF AN INDUSTRIAL LEADER

IT WAS IN DAYTON in 1923 that Charles Allen Thomas settled for his first job out of the Massachusetts Institute of Technology (MIT), joining Kettering and his team as a research chemist at General Motors Research Corporation. Thomas landed the job at GM's Ethyl Corporation fuels division with a helpful reference from the family of MIT friend Henry (Hank) Belin DuPont Jr. The young DuPont's uncle, Pierre S. DuPont, was then president of GM.

Thomas was born in Scott County, Kentucky, on February 15, 1900. His mother, Frances Carrick Thomas, was deeply religious and well educated. His father and namesake was a Christian minister at Newtown Church in Scott County and Broadway Baptist Church in Louisville. The elder Thomas was a talented orator who had arrived in the United States from Australia in 1890 to attend Kentucky University (now known as Transylvania University) and to study at its College of the Bible. He died when his son was six months old.

Thomas was raised by his mother, and the two moved from the Carrick farm near Newtown to the city of Lexington when he was seven years old. They lived at his aunt's home, Hope House, at the corner of Third and Mill streets across from the Transylvania University campus. The location was a fertile academic environment for the aspiring chemist. By the age of 10, he had assembled his first chemistry set and built one of the first wireless stations in Lexington in the days before radio. "Throughout my schooling, I preferred science to any other study," Thomas recalled.[1] Thomas attended Hamilton College's preparatory school in Lexington from the ages of seven to 14. At the age of 13, he took his experiments out of the backyard and into a more formal setting when science professors at Transylvania offered him the use of their laboratories and agreed to mentor him. He completed Morton High School in two years, graduating cum laude in 1916, and continued his studies at Transylvania University.

A WELL-ROUNDED EDUCATION

Outside the classroom, Thomas spent his free time working—helping out on the family farm (1916–20), driving a gas truck, and working in a photo shop. In the summer of 1918, he trained with the U.S. Army in Toledo, Ohio, and was appointed second lieutenant. He returned to Kentucky, joined the Student Army Training Corps, earned the rank of sergeant, and served as a rifle instructor at the Officers Small Arms Firing School at Camp Perry near Lake Erie until the end of 1918. Camp Perry was, and remains, noted for its marksmanship training, a skill that Thomas maintained throughout his life.

Thomas began grooming his social and leadership skills early, and was involved in many activities at Transylvania. Among them, he was a member of Kappa Alpha fraternity, the leadership honor society Lampas, Book and Bones senior society, Skookums Club, Dramatic Club, the literary society Periclea, Morton Club, and the Athletic Council. He was a substitute football player and team manager, and a member and manager of the glee club, as well as a soloist. Additionally, he was advertising manager of the 1920 *Crimson* and was the school's Washington's Birthday speaker in 1919, discoursing on "The Tyranny of the Mob" regarding the menace of Bolshevism. "To have been a boyhood playmate would have been a stimulating experience,"

2.1. Charles Allen Thomas, c. 1922. Courtesy of the Thomas family.

friend Henry DuPont recalled. "Charlie Thomas was no average boy. He was an extrovert type with boundless energy, and very early in life he developed an unusually active mind and a gift for leadership."[2]

After receiving a bachelor's degree in chemistry from Transylvania in 1920, Thomas moved on for graduate studies at MIT. He had considered a career as a professional singer and was intrigued by architecture—an interest he maintained throughout his life—but followed instead his love of chemistry. Though a graduate student, he took practically every undergraduate course in chemistry and was noted for his ability to present complex issues in a simple manner. He earned money to help pay for his studies by singing as a soloist in Boston-area churches and other venues, and also performed with MIT's Technology Glee Club. The school newspaper reported on the young tenor's talents: "The group of songs by C. A. Thomas were delightful, as this young singer's selections always are. Boston audiences are usually enthusiastic over Thomas' voice with its clarity as well as range. He was given a most enthusiastic reception, both because of his versatility and because of the popular nature of the song."[3]

Precocious and confident, Thomas was also a dandy, writing frequently to his mother from MIT about dinners with friends and their parents at spots like the swanky Copley Hotel in Boston, social connections, eligible women, fashion, Harvard-Princeton football games, and visits to the DuPont family estate in Ardmore, outside of Philadelphia. His letters provide a window into his world: He was self-assured, fretted about the debt he was accruing in school and the rigors of completing his thesis, and was self-conscious about going prematurely bald, turning to a product called "Bare to Hair" to invigorate hair growth. He stood 5'10" and was of a slight build at 165 pounds.

GENERAL MOTORS

Thomas aspired to a life as an industry executive, and in order to reach that goal, heeded the advice of his friend Henry DuPont's uncle to begin as a researcher and get a broad foundation. The elder DuPont helped both Thomas and the younger DuPont land a job in

Dayton at GM's Ethyl Corporation fuel division. Thomas was eager to get started on his professional life and began work there in August 1923, before he had completed his MIT degree. He finished his Master of Science degree in chemistry the next year with a thesis on the preparation of benzene sulphonyl chloride.[4] At GM, Thomas worked under Thomas Midgley Jr. and Thomas A. Boyd in the development of tetraethyl lead, an additive for anti-knock gasoline, which had been discovered two years earlier. Though efficient at stopping knock, the gasoline caused problems by coating the interior of engines with lead oxide, which short-circuited spark plugs. Thomas, who was paid $30 weekly, reported on his new job in a letter to his mother:

"Tuesday night"

After the second day of work I'm rather tired—but the work proves very interesting. Every man I've met is a gentleman and very considerate of my "newness" and seems to take the trouble to explain everything. In the first place they all knew who we were and when we would arrive. The plant is over a thousand feet long and proportionately wide—alloy concrete with the walls and some of the ceilings of plate glass windows which open and white canvas shades are drawn over these openings which make the place very comfortable and light. It is four miles in the county from the city limits. The Dayton Wright aeroplane field is across the pike and nearby is the Arkeaydia [sic] Inn, where we live. There are four hundred and sixty some odd men employed—of these only fifty or sixty are research engineers (of which Hank and I are one)—the rest are all expert craftsmen who when a new idea is born, under the directions of the inventor, actually make the thing and thus they can be tried out! When you consider only about one of several hundred attempts on experiments are successful, you realize the expenses of the place—yet General Motors considers this Research Cooperation (as it is called) their biggest asset. One has everything at his command and some of the things they do are remarkable. For instance I saw them testing out the new Cadillac and Buick engines today. To tell you every detail would be interesting of how the temperature of the entering air, water and gas and the temperature of the exhaust etc. are watched—but it takes too long and as I'm sleepy I'll wait and tell you. You understand the motors are mounted on a stand and geared directly to a generator which by watching the variation in the current produced one has a direct means of calculating the efficiency horsepower—for watts can be transformed to horsepower. Another thing, they have a tract of 86 miles of the roughest to the smoothest road. Three crews of men are kept running these new cars day and night (never stopped) in and out until they are literally run to death. Sometimes they go over 50,000 miles before the motor jams—but they never do. These measurements of ware [*sic*]—loses, etc. are noted. Of course all of these tests are more severe than is actually practiced because there the motor runs very intermittently say 8 hours of a maximum.

Well I landed into the department I wanted to get in—the fuel department. I'm working on ethyl gas, which by the way is some dope. I'm under Thomas Midgely Jr. who three months ago got the English "Prix de London" of chemistry for his work on motor fuels. They gave me my choice to go there or in the metallurgical department and excepted [*sic*] both . . . that is I handed this line 'I was out for experience and knowledge and I would like to work in both departments' and they said I could change anytime. I though like the motor fuel work best so I took it first. I have a lab and a man 35 years old to wash my apparatus and run errands. He complains of the cement floor hurting his feet and he talks too much—he don't give me a chance. I've got an interesting problem—the last one so Midgley says from making ethyl gas an ideal fuel.

Well I've got to sleep—we may be down next Sat. as we don't know anyone here.

Lots of love,
Sonny[5]

Thomas networked easily in his new surroundings and was welcomed into Dayton society. The young researcher's social updates to his mother continued from his lodgings at the Moraine Hotel and then at 709 Oakwood Avenue. He reported on dinners at the exclusive Miami Valley Hunt and Polo Club, which he had recently joined. His club membership was discounted to $12 a month upon recommendation by Harold E. Talbott Jr., co-founder of the club, part owner of GM, and a powerful member of the Dayton business and social world. With a membership of about 100 men who kept stables of "ponies," the club provided upscale company for the young engineer. "I am afraid I am way over my head—but what can I do?" he wrote to his mother in May 1924 about his new social life.

The Talbott family legacy in Dayton was substantial and would come to play a central role in both the life of Thomas and that of the atomic bomb project. Harold E. Talbott Jr., with whom Thomas had crossed paths, was the oldest of nine Talbott children born to Harold E. Talbott Sr. and Katharine Houk Talbott. The elder Talbott, who had died in 1921, was one of the founders of Delco (Dayton Engineering Laboratories Co., a subsidiary of GM), Dayton Metal Products Company, and the Dayton-Wright Airplane Company, as well as a partner in all three businesses with Kettering and Orville Wright. A civil engineer, he had built dams, locks, and bridges, and had helped construct the Canadian and U.S. rail systems. His wife, Katharine Houk Talbott, was daughter of Ohio Democratic Representative George Houk and a leader in the arts and civic affairs. Music formed a central part of her life and led her later in life to create the Westminster Choir College. As a clan, the Talbott family was large and active, whether competing in shooting sports, hosting events in their music room, or driven to excel in whatever they attempted.[6] Runnymede Estate, the Talbott home, sat high on a hill in Dayton's affluent Oakwood neighborhood, where Harold Talbott Sr. served as the town's first city manager. Within a year of moving to Dayton, Thomas was invited to the Talbott hunting lodge. "Now socially—boy I am having a whirl. The Talbotts sure are giving me the rush. I am going out to their lodge this weekend to stay until Sunday," he reported to his mother.[7]

At work, Thomas partnered with Dayton native Carroll A. (Ted) Hochwalt, a University of Dayton chemical engineering graduate. Hochwalt had been on vacation when Thomas arrived at GM, and returned to find "a guy with a red fringe of hair around a bald head, who was singing heartily away" in the room next door.[8] The two personalities complemented each other. Hochwalt was calm, studious, and soft-spoken. Thomas was gregarious, articulate, and charming. Hochwalt described Thomas as possessing "imagination, curiosity, intuitiveness, buoyant enthusiasm, driving ambition, friendliness of a cheerfully extroverted type, genuine interest in his friends and associates, and an apparently unlimited supply of energy."[9] Both men were ambitious, confident, and had a visionary and pioneering ability to predict new directions for science and engineering.

Hochwalt was in charge of developing the synthetic process and chemistry of tetraethyl lead; Thomas was tasked with finding a "scavenger" to remove lead deposits from engines. The two men discovered that adding ethylene dibromide to the anti-knock fuel scavenged the lead and prevented it from sticking to engines. At the time, however, Germany controlled the world's supply of bromine, and the price was prohibitively high for use in consumer products. The solution proved to be "mining" seawater, which contained bromine and sodium. The Tribromoaniline Method of obtaining bromine from seawater involved adding chlorine to seawater to form free bromine and sodium chloride or salt and then extracting the bromine with an organic compound that with a frothing agent could be raised to the surface of water. To determine which part of the Atlantic contained the most bromine, Thomas got in a rowboat at various points up and down the coast and collected samples. The Ethyl Corporation that employed Thomas had been formed in 1923 by a partnership between GM and Standard Oil of New Jersey to pursue application of GM's tetraethyl lead patent. Du Pont was hired to produce tetraethyl lead and created a floating pilot plant to test the process by converting a cargo steamer that cruised the Atlantic Coast sucking in seawater and producing bromine. Thomas described the ship, which had been christened *Ethyl*, as the most inefficient chemical plant ever.

PRODUCTIVE PARTNERSHIPS

The Thomas–Hochwalt creative team had no shortage of ideas. Early in 1925, the men approached Kettering and asked him if they could work in the evenings on their own projects. They got the go-ahead, sometimes using their own lab, sometimes that of Hochwalt's mentor, William Wohlleben, founder of the University of Dayton's chemistry and chemical engineering departments. In 1926, when Ethyl Corp. announced that the Dayton fuel research laboratory would be moved to Detroit and consolidated with GM's General Research Department, Thomas, Hochwalt, and Midgley declined to move. Among

2.2. Runnymede Playhouse was the largest free-standing private hall in the country when it was built in 1927. Courtesy of the Talbott family.

the reasons Thomas wanted to stay in Dayton was his interest in the youngest of the Talbott children, Margaret Stoddard Talbott, known as "Marnie." The two were married at Runnymede on September 25, 1926. Marnie was 20; Charlie was 26.

The match between Charlie Thomas and Marnie Talbott was a powerful and often competitive one. Theirs was a social, athletic, creative, and ambitious world. Marnie was a member of the All-American skeet team and a natural competitor in everything from card games to conversation. Thomas loved the spotlight. They were both adept at socializing and moved in the upper rungs of Dayton society. The newlyweds made their home at 236 Rubicon Road in Dayton and, within a year of marriage, their first child was born, Charles Allen Thomas III. Three daughters followed: Margaret, Frances, and Katharine. The young family spent a lot of time at Runnymede with the Talbott clan, which included Marnie's eight siblings and their numerous offspring. To better host family events, matriarch Katharine Talbott built a recreation center adjacent to her Runnymede

2.3. The Playhouse was frequently used for community events, including bridge parties, plays, and concerts. Courtesy of the Talbott family.

home in 1927. Known as "Runnymede Playhouse," the glass-roofed structure was the largest free-standing private hall in the nation at the time and was used by the Talbotts and the Oakwood community for functions ranging from Talbott Sunday luncheons and skit nights to school plays, community recitals, and exhibition tennis matches.

The building, constructed at a cost of $100,000, housed a stage, dressing rooms with Italian marble showers, a tennis court with a green cork floor, a squash court, a kitchen, and a greenhouse full of tropical plants. The second floor contained a lounge with marble counters, fireplace, and two tiers of balconies that overlooked the main floor. Outdoors, a swimming pool and cobblestone courtyard completed the facility, which was entered through brick gates.[10] When large family gatherings called for it, as many as 90 Talbotts would dine on the main floor. It was also large enough to accommodate 1,200 women for an afternoon of bridge; an exhibition tennis match by local tennis star Virginia Hollinger and the first American to win the Wimbledon tennis championship, Bill Tilden; and annual Christmas

2.4. The Talbott clan gathers in the Playhouse for Christmas dinner, c. 1930. Courtesy of the Talbott family.

carol programs by students at the local Harman Elementary School. Although Katharine Houk Talbott died in 1935 and her house was torn down two years later, the Runnymede Playhouse remained and continued to be used by the family and community. Thomas was a regular on the squash court during the winter months.

CHAPTER 3

THOMAS & HOCHWALT
LABORATORIES AND MONSANTO

WITH HIS PERSONAL LIFE settled in 1926 by marriage, Thomas focused his energy on growing the laboratory business that he and Hochwalt established on the third floor of a former residence known as Grey Manor Annex at 127 North Ludlow Street in downtown Dayton. The building was owned by the Talbotts and included a public tearoom on the first floor that was sometimes in the path of experiments gone awry. A family history contains the following account:

> Almost immediately a minor experiment of theirs blew a minor hole through the roof; that was patched up and the research continued. Then one day when (Mrs. Talbott) was in the Annex she noticed smoke curling up between the floorboards under her feet. She pretended not to have noticed, but quietly asked someone to investigate. Charlie and Ted had perfected a new type of fire extinguisher and had built a fire in the cellar to extinguish.[1]

Thomas and Hochwalt had begun the fire extinguisher work while moonlighting in labs at the University of Dayton. Unlike the

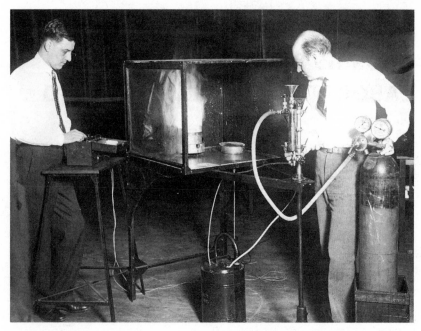

3.1. Chemists Carroll Hochwalt, left, and Charles Allen Thomas conduct research on their fire extinguisher. Courtesy of the Thomas family.

soda-acid extinguishers of the day, which froze unless kept indoors, their fire extinguisher was revolutionary because it could function in temperatures as frigid as 40-below and could be used to fight outdoor fires such as those in lumber yards.[2] It used alkaline metal salts mixed with water to create a foam that was effective on oil fires.

The findings were reported in the team's first publication, "Effect of Alkali Metal Compounds on Combustion," which was presented at the 1928 American Chemical Society meeting in St. Louis, Missouri (April 16–19, 1928).[3] *Chemical & Engineering News* later reported on the event: "The address (at the St. Louis meeting) called for a demonstration and, naturally, the demonstration called for the use of burning gasoline. At this point the manager of the Chase Hotel where the session was being held insisted that the program committee head give him a guarantee that the hotel would not burn down."[4] As a precaution, the stage was lined with fire extinguishers. The demonstration was a success—the hotel did not go down in flames—and the invention was sold to the FyrFyter Company, where it became a top seller.

Thomas & Hochwalt Laboratories received royalties on the product for 17 years.

Lab staff initially included the two chemists, two lab assistants, and a secretary who was paid $18 a week.[5] Despite the duo's split from General Motors, relations between Thomas, Hochwalt, and their former employer were good. GM was the lab's first client, paying $150,000 a year for lab space for Thomas Midgley and for work on a synthetic substitute for rubber.[6] The income helped subsidize the business for its first two years.

In 1928, the partners incorporated as Thomas & Hochwalt Laboratories, deciding leadership by the flip of a coin—Thomas as president and Hochwalt as vice president. They were joined on the board of directors by Thomas's brothers-in-law, Harold E. Talbott Jr. and Nelson S. Talbott, who provided capital and business acumen. The lab's research staff included eight employees, counting the two founders. Thomas, with his extroverted personality, was the natural salesman. "(Thomas) and Hochwalt used to spend long evenings figuring and plotting what they would do if they were in charge of research for various corporations. Then Thomas would seek audience with the president of the lucky company they had chosen to favor with their talents and proceed to outline a research program for the company. Strangely enough it worked," *Monsanto* magazine later reported.[7]

One of the lab's early contracts was with Alemite Corporation for a carbon-removing agent used in automobile engines and marketed as Carbosolve by Chrysler. Consumers emptied a can of Carbosolve into their car engines and let it stand overnight. The next morning, when the car was started, the carbon was blown out.[8] For this project, as with others, Thomas & Hochwalt Laboratories charged a lump sum. The partners also demanded royalties from all patents developed. The arrangement was cost-effective; Carbosolve netted the partners $300,000, which they used to purchase the former Dayton Chemical Company building at 1515 Nicholas Road, about three miles south of downtown Dayton on the Miami River. The partners had only recently moved into the newly renovated facility in March 1929 when a hydrocarbon cracking experiment exploded, destroying a $7,000 library and all laboratory equipment. There were no injuries, but the facility was sufficiently damaged that activities had to be moved to the University of Dayton while the lab was rebuilt.[9] Explosions aside,

Thomas & Hochwalt Laboratories was beginning to draw serious attention, especially from the Monsanto Chemical Company, which was supplying the business with greater and greater quantities of chemicals. Monsanto business development executive Lynn A. Watt was sent from the company's St. Louis headquarters to investigate.[10] After the visit, Thomas & Hochwalt Laboratories began handling regular research contracts for Monsanto, primarily in the phosphorous and detergent group. Projects for other clients included sizing agents for paper and the preparation of synthetic resins from petroleum derivatives. Standard of New Jersey took a $50,000 option on the resin project but backed out and was replaced in 1933 by Monsanto, which purchased a major share in the project and formed the subsidiary Monsanto Petroleum Chemicals Inc.[11] Monsanto marketed one of the resin products as Santolube for use in engine crank case oils to protect alloy bearings against corrosion and stabilize engine oil against heat.[12]

During the early years of the lab, Thomas also pursued his own interest in a fundamental theory on aluminum chloride, which makes it possible for a number of compounds to react with benzene, one of the basic building blocks of synthetic chemicals. From that work came a 1941 American Chemical Society monograph *Anhydrous Aluminum Chloride in Organic Chemistry*. The book served as a standard reference work for many years for those working with aluminum chloride. With his foot firmly planted on the ground of industrial chemistry, Thomas also organized the American Chemical Society's (ACS) Dayton Section in 1931, eventually becoming national ACS board chairman in 1948.

Chemistry was both a business and a fascination for Thomas. In addition to standard chemical contracts, Thomas & Hochwalt Laboratories was subject to Thomas's creative whims. He loved potato chips and was determined to create an automated potato chip machine that would shorten prep time. He patented a device that sliced, washed, dried and cooked chips. Among the lab's clients were nationally known magicians Harry Blackstone and Howard Thurston. Thomas concocted a trick for Blackstone in which the magician showed his spellbound audience that he could walk away from his shadow (using fluorescent zinc sulfide screen). Thomas, himself, was an occasional entertainer in alchemy and used the trick to entertain at social and business gatherings.[13] His endeavors, both straight-line

3.2. Though a science administrator and industrial chemist for Monsanto, Charles
Allen Thomas pursued his own interest in a fundamental theory on aluminum
chloride, which makes it possible for a number of compounds to react with
benzene, one of the basic building blocks of synthetic chemicals. From that
work came a 1941 American Chemical Society monograph, *Anhydrous Aluminum
Chloride in Organic Chemistry.* Courtesy of the Thomas family.

science and those beyond mainstream, eventually resulted in nearly 100 patents. "Thomas was always coming in with wild ideas. Hochwalt was the restraining one," longtime lab secretary Ruth Dornbusch Setzer recalled.[14] Among the less-traditional projects were:

- A method of preserving cut flowers that for a time filled the lab with roses every day.
- An unspecified project that involved boiling shoe tongues, a process that reportedly led to bad odors.
- Research on self-rising flour that amused his children as the experiments continued in the home kitchen.[15]

SURVIVING THE DEPRESSION

With the onset of the Depression, business declined at Thomas & Hochwalt Laboratories. The partners' investment in the land and facilities on Nicholas Road just before the October 1929 Crash stood them well, but contracts were down. Of those that remained, clients included Morton Salt, for which they developed a synthetic salt for curing hams, and Chrysler, likely maintained by Talbott family ties. Money was allegedly so tight that when President Roosevelt closed the banks, payroll was met through the partners' winnings in a craps game.

The Depression spurred innovation in the lab, including a method for analyzing bootleg liquor to help clients avoid toxic and potentially lethal moonshine. "A man would have a case to be analyzed and Midgley would get a pint to analyze and then ask for two quarts in payment. His way of testing it was to drink the pint sample, and if nothing happened, that was okay. That man could drink anything," Thomas recalled.[16]

A more legitimate exercise in alcohol chemistry involved a method of quick-aging whiskey developed just before the 1932 presidential election. Thomas and Hochwalt predicted a victory by Franklin D. Roosevelt and the subsequent repeal of Prohibition. They knew that distillers didn't have enough aged alcohol stock on hand to meet consumer demand, so six to eight months before the election, they began researching ways to artificially age raw, green liquor. They signed a contract with National Distillers and developed a pilot plant at Louisville's Old Granddad Distillery.[17] When Prohibition was repealed in 1933, three million bottles of bourbon were ready to go on the shelf. The lab was paid $250,000 for the job. "We figured out a way to beat the clock and to bring about rapid aging. It wasn't very good bourbon. But it was pure, wet, alcoholic and available," Hochwalt reported.[18]

With a family to support and money tight, Thomas also sought outside work during the Depression and joined Sharples Solvents in 1932 as research director. Thomas & Hochwalt Laboratories had done contract work for the company on the problem of "orange peel" wrinkling of lacquers; the firm was owned by friend Phil Sharples. The research involved amyl mercaptans (a powerful odorant that smelled like rotten eggs), which got into the Dayton sewer system and drew complaints from residents that it was seeping out into homes: "The police came around and said it had been detected as far away as Springfield. The Board of Health came out and said they would have to stop it," lab secretary Ruth Setzer recalled.[19] In order to take the position with Sharples, Thomas relocated his family to Philadelphia, living in a house on Spring House Lane in Wayne. It was not a happy time for the chemist, who wrote to his mother that he felt like he was being drawn away from research:

Dearest Mother:

I'm writing this in the laboratory, a place where I love to be but it seems that I'm always being pulled out most of the time. Last Sunday was the first one in six weeks that someone hasn't been out on business or something about work—true it's only for 3 or 4 hours of the day but I'm mentally tired of business. My mind relaxes under a scientific problem and I'm eager for it—but negotiations are nerve racking to me now. In self analysis I've come to the following conclusions. I've some talents along too many lines. At heart I'm a pure scientist (not that I'd ever be a great one), but I'd like to leave its opportunity of finding out what I could do but it seems that for negotiating a business deal and executive work I always find myself with its responsibility. I've had a good deal of success with the latter but to do better is impossible physically and mentally and I know it—but I keep on doing it. Perhaps my analysis is wrong and I should do nothing but the latter. Of course the Depression has caused some of my mix up but I'm not happy now under this strain.

I feel that I must think things through and decide my own life and
have enough guts to tell anyone that money or no money I'm going to
have fun with the work I want to do.[20]

LIFE OUTSIDE THE LAB

In 1934, when contracts with Monsanto in the phosphorus and coal
tar chemical area began to pick up, Thomas returned to Dayton. The
family settled 10 miles southwest of Dayton off Mad River Road on
a 12-acre property they named Blue Grass in a nod to Thomas's Kentucky roots. It was a gentleman's farm, chemistry laboratory, engineering pilot plant, and recreation center all in one. At one point, Thomas
added a flock of sheep to serve as lawn mowers and co-owned some
Jersey dairy cows with his next-door-neighbor and business partner,
Hochwalt. Thomas used the property to experiment with his lifelong love of farms and fascination with engineering, an interest that
later translated into Monsanto's expansion into agricultural activities. A patio and back terrace were paved with old lithograph stones
from Bavaria and Cincinnati depicting letters and images. Thomas
designed a system to filter iron out of the well water and engineered a
graded sand filter system to provide clear water to the swimming pool.
He also designed an early form of motion-detection driveway lighting
that was triggered by the headlights of incoming cars. The lights fell on
a small window in a shingled weatherproof box, behind which was a
selenium light-sensitive vacuum tube. The box contained the rest of the
electronics and relays that when hit by the headlights of an approaching car switched on about six lights leading up the hill to the house.

Thomas was intrigued by plumbing, advising his young son to
"learn how to plumb pipes and you'll never go hungry." To reinforce the lesson, he directed his son to paint the exposed pipes in the
basement of the main house, color-coding them for function: blue for
incoming well water, red for hot water for heating the house, yellow
for returns, black for gas. The basement looked like a chemical plant,
and Thomas showed it off proudly to visitors and dinner guests. Also
in the basement was a chemistry laboratory replete with gas-fed Bun-
sen burners that his son, who later received a doctorate in chemistry
from Harvard, used for amateur experiments.

In his down time, Thomas and his wife frequented the Miami Valley Hunt and Polo Club, which had skeet fields, tennis courts, a swimming pool, a polo field, and an active social scene. The couple attended parties there on summer evenings, during which Thomas often sang, accompanied by Marnie who sang alto. In the winter, Thomas and his wife played fast-paced and competitive badminton games at the Moraine Country Club. He also belonged to Dayton's Engineers Club for professional networking and the Buz Fuz men's club for prominent Dayton citizens.[21] Later in life, family musicales were routine for the Thomas family, with his youngest daughter, Katharine—a trained singer—frequently joining him in the spotlight. "At award dinners and less formal meetings Dr. Thomas is almost invariably called upon for a rendition of 'Wagon Wheels'—his tenor voice is almost as well-known as are some of his outstanding scientific achievements," a reporter wrote.[22] A natural-born entertainer, Thomas loved the attention.[23] It was from this world that Thomas commuted to his Nicholas Road laboratory, driving the rural roads in his two-seat Plymouth roadster. The distance gave him time to clear his head and consider matters. His Schnauzer, Bonco, sometimes accompanied him on the drive and spent the day tied to Thomas's desk; his secretary walked the dog. Thomas worked almost exclusively at the lab, and there were few professional artifacts or work files around the house. Some weekends, his young son accompanied Thomas to the lab and was let loose to play in empty laboratories where he conducted his own experiments.[24] This amateur work didn't always meet with approval from Thomas's colleagues, it was reported:

> Charlie's little boy used to come in on weekends with his father and fool around while Charlie was working. He would always take over the laboratory of a man named Minton. The kid used to mix up stink bombs and make a big mess of the place. Minton came to dislike him intensely.[25]

MERGING WITH MONSANTO

In the spring of 1935, Thomas and Hochwalt cast their net nationally and paid $3,000 for a full-page ad in *Fortune* magazine that described their laboratory as "a unique organization in chemical science, which serves as consultants to many of America's largest industrial offices and laboratories."[26]

> Day by day, chemical research changes the world in which we live.
> New discoveries open new roads for business enterprise . . . new work
> is found for men, and, in the light of chemical research . . . brighter
> grows the way.[27]

The investment paid off. By the end of 1935, Thomas & Hochwalt Laboratories—with about 80 chemists and engineers—had become the largest independent consulting laboratory in the country. Its largest client was Monsanto, which by then was contributing about 30 percent of the lab's contracts. The laboratory's work, and the growing size of the business, drew the attention of Monsanto president Edgar Queeny, who in January 1936 made a sudden offer for the lab, declaring that he wanted to enhance his company's scientific reputation and expand its scope of activities. During a visit to Dayton, he told Thomas and Hochwalt, "Your bills are so high it would be cheaper for Monsanto to buy you out."[28] In reality, many thought that Queeny, who "bought talent along with companies . . . absorbed Dayton's Thomas and Hochwalt Laboratories mainly to snare Charles Allen Thomas and Carroll A. Hochwalt."[29] The deal, worked out in a month of negotiations, closed on February 25, 1936, Easter Sunday. Monsanto purchased Thomas & Hochwalt Laboratories for $1.4 million in common shares of Monsanto stock. Queeny reportedly thought the price was high, to which Thomas responded, "A stable of thoroughbreds always comes high."[30] The merger allowed Monsanto to create a long-range research program and established the chemical company's Central Research Department. It was first known as Thomas & Hochwalt Laboratories Department of Monsanto, later as Monsanto Central Research Division, and then as the Monsanto Dayton Laboratory. By 1983, Monsanto had phased out the work in Ohio and transferred all research to its St. Louis headquarters.

With the sale of the laboratory, Thomas was named director of Central Research in Monsanto's St. Louis headquarters, then an eight-hour train ride from Dayton; Hochwalt replaced him as director of the Dayton facility. It was the beginning of a long relationship

3.3. Thomas & Hochwalt Laboratories on Nicholas Road in Dayton, following the merger with Monsanto Chemical Company in 1936. Courtesy of the Thomas family.

with Monsanto for the two men. Thomas became Monsanto president in 1951, then board chairman before retiring in 1970. Hochwalt led Monsanto's Dayton lab from 1945 to 1951, at which time he was transferred to Monsanto's St. Louis headquarters. He retired in 1964. In the first few years after the merger, personnel at the Dayton facility increased by about 15 percent and a building expansion was designed by Dayton architect Douglas Lorenz.[31] When complete, the 12,200-foot addition included 14 laboratories, an instrument room, stock room, sample room, washroom for apparatus, office for the chemists, locker room, kitchen, and cafeteria. Most of the new employees were PhDs recruited from Harvard, MIT, Pennsylvania State University, the University of Illinois, University of Michigan, University of Chicago, and the University of Dayton. Among the contacts who helped refer talent was Harvard University President James B. Conant.

Monsanto's early work in Dayton focused on surface-active agents (detergents for removing dirt and stains from fabrics) and phosphate builders (for keeping the dirt in wash water suspended). The research led Monsanto to become a major supplier in the detergent industry and—in the 1940s—to the invention of the first low-sudsing detergent known as All.[32] "Much of Monsanto's proprietary and patented chemical knowledge traces back to Dayton, but even more to the succession of researchers nurtured in the spirit of Thomas and Hochwalt," wrote Monsanto corporate historian Dan J. Forrestal.[33] The two men worked in the areas of petrochemicals, detergents, and fibers—most notably on styrene.[34] Among the projects was work begun in 1938 on styrene monomer, which was then viewed as a raw material of potential use in plastics.[35] As part of this research, Monsanto built a styrene pilot plant in Dayton.

Shortly after the Monsanto purchase, Thomas and his wife traveled to Germany to visit chemical engineering operations there and meet with top German scientists. This was three years before the outbreak of World War II, and the United States was still doing business with Nazi Germany. With Germany's invasion of Poland on September 1, 1939, Monsanto's central laboratory and the world experienced a change in direction. The war brought extra work to American industry, including Monsanto's Central Research Department. It surely cast a new light on the high-placed Nazi scientists Thomas had met in Germany.

CHAPTER 4

U.S. SCIENCE AND INDUSTRY PREPARE FOR WAR

IN DECEMBER 1938, chemists Otto Hahn and Fritz Strassmann, working with physicist Lise Meitner, found evidence of nuclear fission in uranium, which unlocked the potential of atomic weapons. The three researchers had worked together at the Kaiser Wilhelm Institute in Berlin; Meitner, a Jew, fled Nazi Germany to Holland in July 1938 and then to Sweden but continued collaborating with the others. The team discovered that a uranium atom bombarded by neutrons splits into two lighter nuclei and releases energy, while also creating two or three additional neutrons. In Sweden, Meitner and her nephew, physicist Otto Frisch, further studied the problem, concluding that a new process, fission, was in place and could release a tremendous amount of energy.

Frisch discussed the fission findings with Danish scientist Niels Bohr, who brought the information to the U.S. with him when he arrived in January 1939 to work at the Institute for Advanced Study in Princeton, New Jersey. Bohr shared the news during a physics conference at George Washington University in Washington, DC. The announcement of the discovery alarmed U.S. scientists who had

4.1. Future Manhattan Project leaders meet at the University of California, Berkeley, in March 1940 to discuss physicist Ernest Lawrence's 184-inch cyclotron. Left to right: Ernest Orlando Lawrence, Arthur H. Compton, Vannevar Bush, James B. Conant, Karl T. Compton, and Alfred Loomis. Photo courtesy of Lawrence Berkeley National Laboratory. © 2010 The Regents of the University of California, Lawrence Berkeley National Laboratory.

been working on solving the fission problem and feared that the U.S. government's failure to support its own uranium research would result in Germany being first to develop and use an atomic bomb. On July 12, 1939, physicists Leo Szilárd and Eugene Wigner, representing American scientists who were researching uranium and plutonium, approached Nobel laureate Albert Einstein at his cottage on Long Island. They asked him to lend his name to their effort to get government support for the research by writing to President Roosevelt, warning him of the possible use of a nuclear chain reaction in a German bomb and urging him to support uranium weapons research. Einstein wrote on August 2, 1939: "Some recent work by E. Fermi and L. Szilard, which has been communicated to me in manuscript, leads me to expect that the element uranium may be turned into a new and important source of energy in the immediate future. Certain aspects of the situation which has arisen seem to call for watchfulness and, if necessary, quick action on the part of the Administration . . . In view of the situation you may think it desirable to have more permanent contact maintained between the Administration and the group of physicists working on chain reactions in America."[1] Economist Alexander Sachs, who had a personal connection to the president, took the letter to President Roosevelt on October 11, 1939. In response, Roosevelt created an advisory committee led by Lyman Briggs, director of the National Bureau of Standards, and including representatives of the Army and the Navy. The Advisory Committee on Uranium met for the first time on October 21, 1939, and recommended that the government fund fission chain reaction research. The committee did not indicate, however, that any great rush needed to occur, as the country wasn't under threat of war. The funding it offered was minimal—including $6,000 for physicist Enrico Fermi's work at Columbia University on neutron absorption in graphite.[2]

SCIENTISTS AND GOVERNMENT ORGANIZE

From the creation of the Uranium Committee to the ultimate designation of the Manhattan Engineer District of the U.S. Army Corps of Engineers (popularly known as the Manhattan Project), administration of the bomb project took many turns. Beginning with the formation of the Uranium Committee, scientists, the government, and the military partnered on the effort in a variety of forms until the ultimate atomic bomb organization emerged. The chronology and details of the work by scientists and committees is presented in a greatly condensed form to introduce key committees and administrators.

NATIONAL DEFENSE RESEARCH COMMITTEE

In June 1940, with little action from the Uranium Committee and pressure from Carnegie Institution President Vannevar Bush, President Roosevelt authorized the creation of the National Defense Research Committee (NDRC), which was "to coordinate, supervise, and conduct scientific research on the problems underlying the development, production, and use of mechanisms and devices of warfare."[3] The

Advisory Committee on Uranium became a NDRC subcommittee. Bush was appointed chair and was joined by the following committee members:

> Richard C. Tolman, dean of the graduate school and professor of physical chemistry and mathematical physics at the California Institute of Technology
>
> Irvin Stewart, former Federal Commissioner for Communications and chairman of the Committee on Scientific Aids to Learning
>
> Rear Admiral Harold G. Bowen, director of the Naval Research Laboratory
>
> Conway P. Coe, U.S. Commissioner of Patents
>
> Karl T. Compton, president of the Massachusetts Institute of Technology
>
> James B. Conant, president of Harvard University and a chemist
>
> Frank B. Jewett, president of the Bell Telephone Laboratories and president of the National Academy of Sciences
>
> Brigadier General George V. Strong

The NDRC was divided into five research divisions—and sections within those divisions:[4]

Division A (Armor and Ordnance)

Richard C. Tolman, chairman; Charles C. Lauritsen, vice-chairman
- Section B (Structural Defense)
- Section H (Investigations on Propulsion)
- Section S (Terminal Ballistics)
- Section T (Proximity Fuses for Shells)
- Section E (Fuses and Guided Projectiles)

Division B (Bombs, Fuels, Gases, Chemical Problems)

James B. Conant, chairman
Synthetic Problems—Roger Adams, vice-chairman
- Section A-1 (Explosives)
- Section A-2 (Synthetic Organics)
- Section A-3 (Detection of Persistent Agents)
- Section A-4 (Toxicity)

Physical Chemical Problems—Warren K. Lewis, vice-chairman
- Section L-1 (Aerosols)
- Section L-2 (Protective Coatings)

- Section L-3 (Special Inorganic Problems)
- Section L-4 (Nitrocellulose)
- Section L-5 (Paint Removers)
- Section L-6 (Higher Oxides)
- Section L-7 (Oxygen Storage)
- Section L-8 (Gas Drying)
- Section L-9 (Metallurgical Problems)
- Section L-10 (Exhaust Disposal)
- Section L-11 (Absorbents)
- Section L-12 (Oxygen for Airplanes)
- Section L-13 (Hydraulic Fluids)

Miscellaneous Chemical Problems

- Section C-1 (Automotive Fuels; Special Problems)
- Section C-2 (Pyrotechnics)
- Section C-3 (Special Problems)

Division C (Communication and Transportation)

Frank B. Jewett, chairman; C. B. Jolliffe, Hartley Rowe, R. D. Booth,
and J. T. Tate, vice-chairmen

- Section C-1 (Communications)
- Section C-2 (Transportation)
- Section C-3 (Mechanical and Electrical Equipment)
- Section C-4 (Submarine Studies)
- Section C-5 (Sound Sources)

Division D (Detection, Controls, and Instruments)

Karl Compton, chairman; Alfred L. Loomis, vice-chairman

- Section D-1 (Detection)
- Section D-2 (Controls)
- Section D-3 (Instruments)
- Section D-4 (Heat Radiation)

Division E (Patents and Inventions)

Conway P. Coe, chairman

Thomas, who was known through industrial and scientific
circles to both Conant and Compton, was brought into the NDRC
as a special investigator and later assigned as a consultant to Division 8, a further specialized division that was led by Harvard chemist

George Kistiakowsky and was concerned with research on explosives. Between October 1940 and April 1941, Thomas led liquid fuels research for the NDRC.

OFFICE OF SCIENTIFIC RESEARCH AND DEVELOPMENT

By the spring of 1941, Bush had grown concerned that the work of the NDRC's Uranium Committee was progressing too slowly; the British were further ahead on research, and it appeared that the Germans were actively pursuing an atomic bomb.[5] He asked Jewett, president of the National Academy of Sciences, to convene a group to review the matter. That committee, led by physicist Arthur Compton, dean of science at the University of Chicago, issued three reports between May and November 1941 that advised intensive effort on uranium research, and the development of an intermediate-scale uranium/ graphite experiment and a pilot plant for producing heavy water, support for investigating the properties of beryllium as a moderating agent, and ongoing work on isotope separation.[6]

On June 28, 1941, after continued pressure from Bush to increase the NDRC's authority to include engineering development in the military planning of weapons research and permit it to build prototypes of new devices, Roosevelt authorized the creation of the Office of Scientific Research and Development (OSRD). Housed within the Office for Emergency Management, the OSRD was to coordinate nuclear fission research and had the authority to enter into contracts and agreements for experimental study related to weapons research and development. Bush was appointed director of the new organization, with Conant overseeing the NDRC.[7] The NDRC was expanded to include:

Division 1—Ballistic Research

Division 2—Effects of Impact and Explosion

Division 3—Rocket Ordnance

Division 4—Ordnance Accessories

Division 5—New Missiles

Division 6—Sub-surface Warfare

Division 7—Fire Control

Division 8—Explosives

Division 9—Chemistry

Division 10—Absorbents and Aerosols

Division 11—Chemical Engineering

Division 12—Transportation

Division 13—Electrical Communication

Division 14—Radar

Division 15—Radio Coordination

Division 16—Optics and Camouflage

Division 17—Physics

Division 18—War Metallurgy

Division 19—Miscellaneous

MONSANTO'S DEFENSE CONTRACTS

Thomas's star had been rising steadily in the NDRC during this time. In addition to explosives research, Monsanto had undertaken wartime work on another strategic problem—finding a synthetic substitute for natural rubber, which was controlled by the Japanese. Some 90 percent of the natural rubber supply was cut off, with no industrial supply of synthetic substitutes to meet the nation's 600,000-ton per year consumption.[8] Rubber was needed for aircraft, tanks, battleships, footwear, tires, and other critical war equipment; finding alternate sources for rubber was a national priority. By war's end, the United States had spent as much on its rubber program as it did on the atomic bomb.[9]

In June 1940, President Roosevelt had created the Rubber Reserve Company to coordinate the use of all major types of synthetic rubbers: GR-S (government rubber-styrene) copolymers, butyl rubber, and neoprene. Monsanto was well positioned to help, given its operation of a styrene pilot plant in Dayton and purchase in 1929 of Rubber Service Laboratories Co. of Akron, Ohio, and Nitro of West Virginia to develop plastics.[10] By 1941, Monsanto had completed fundamentals for a successful styrene process and initiated production studies at its pilot plant. The company's annual report that year made no mention of the government work, describing the project as "an important raw material for synthetic rubber." A year later, the rubber crisis had become more severe. President Roosevelt appointed a Rubber Survey Committee led by financier Bernard M. Baruch that was charged with mobilizing 51 plants, 20,000 workers, and over half-a-billion-dollars worth of machinery to produce the monomers and polymers needed for the manufacture of synthetic rubber.[11] The committee included Conant, Compton, and Thomas, and enlisted the services of Dow Chemical Co., Union Carbide, and Monsanto, among other companies.[12] In March 1942, Monsanto began construction of a full-scale production unit at Texas City, Texas, and conducted a six-month training program in Dayton for plant operators.

Monsanto's Dayton lab, with a staff of about 100, was also busy with other NDRC contracts, including a project conducted from October 1940 to April 1941 developing a synthetic resin for the British to use to booby-trap gasoline in the event of an invasion by Germany. The resin could be added to gasoline, polluting engines beyond repair. An additional project, initiated four months before Pearl Harbor, led to the development of a new type of composite solid fuel for use in rockets.[13] The resin was produced in a semiworks in Dayton and was planned on an industrial scale at an $8 million plant that would cover 2,000 acres in Marshall, Texas.[14] Monsanto relinquished its patent rights to jet propellants—and other war contracts—as a "contribution to the war effort."[15] Kistiakowsky responded to this corporate declaration, "It seems to me that it sets a magnificent example to American industry in emphasizing that the work done with NDRC funds is work done for our country and not for private interests."[16] The war ended before the fuel plant was in full production, but the project drew praise from OSRD Director Vannevar Bush:

> Dear Dr. Thomas:
> It can now be told, within our own group, that new devices developed through the close collaboration between the services and NDRC, have recently been used in combat with the enemy and have not been found wanting. . . . We seem to have reached the turning point of the war, but the end cannot be predicted. We are in a struggle with a resourceful and ruthless enemy. . . . My congratulations to all who labor to make the American scientific effort thorough, well rounded, effective, and rapid.[17]

In July 1941, NDRC leaders Bush and Conant received a draft report from the British MAUD Committee, their counterpart across the Atlantic.[18] The report maintained that a sufficiently purified critical mass of uranium-235 could fission even with fast neutrons. Based on atomic bomb research by refugee physicists Rudolf Peierls and Otto Frisch, the MAUD report estimated that a critical mass of 10 kilograms would be large enough to produce an enormous explosion.[19] A bomb that size, the report concluded, could be ready in approximately two years and could be delivered by aircraft. The report accelerated the U.S. bomb effort. British Prime Minister Winston Churchill and members of the MAUD Committee felt it was time to construct a bomb. Roosevelt and Churchill communicated about the joint development of a bomb, with possible sites including Canada, the British Isles, or the United States. Locating the project in Great Britain would put efforts too close to enemy eyes and would require too much diversion of manpower from the British war effort.[20] It had become imperative the United States begin bomb production on a centralized, industrial scale. In December, after meeting with Bush, President Roosevelt ordered the formation of a Top Policy Group to handle secure discussion of the project. It consisted of himself, Vice President Henry A. Wallace, Secretary of War Henry Stimson, Army Chief of Staff General George C. Marshall, Bush, and Conant. All policy issues on fission weapons were to funnel through Conant and Bush. Though Bush and the OSRD did not have final permission to build a bomb, they did now have permission to expedite research and planning on the feasibility of one.[21]

The NAS review committee's third report, delivered to President Roosevelt on November 27, 1941, pushed the work further ahead. "A fission bomb of superlative destructive power will result from bringing quickly together a sufficient mass of element U235," it stated.[22] Roosevelt's simple "OK" response on January 19, 1942, moved the project forward.

CHAPTER 5

BIRTH OF THE MANHATTAN ENGINEER DISTRICT

THOMAS FREQUENTLY SPENT weekends at the office. On Sunday, December 7, 1941, as he and his 14-year-old son were leaving the lab, a security guard informed them that the Japanese had bombed Pearl Harbor. "I'll be damned," Thomas responded.[1] Neither Monsanto nor Thomas was connected to the uranium and fission studies then under way in labs around the country, but both were now closer to being involved in the bomb project. Thomas was by then deputy chief of the NDRC explosives division (Division 8) and chief of Section 8.1. The division, in which his business partner Hochwalt was also a member, was responsible for subcontracting with the chemical industry and universities.[2]

As 1942 began, Thomas and the researchers in Dayton were focused on Monsanto's NDRC contracts. Up the road, physicist Arthur Compton was busy at the University of Chicago, organizing the new Metallurgical Laboratory for the bomb project. The lab, which operated under contract with the OSRD, consolidated researchers from around the country to develop chain reacting piles for plutonium production, create methods for extracting plutonium from irradiated uranium, and focus on weapon design. Those moving to Chicago included plutonium co-discoverer Glenn Seaborg from Berkeley and physicist Enrico Fermi from Columbia University.

By the summer of 1942, plans for the bomb program were nearly in place. On June 17, Bush presented a report to President Roosevelt detailing plans for the expansion of the atomic bomb project and dividing the work between the OSRD and the U.S. Army Corps of Engineers. As the exact path to a bomb was still an unknown, the report recommended construction of five plants: a centrifuge plant for production of enriched uranium; a gaseous diffusion pilot plant for uranium separation; an electromagnetic plant for the production of enriched uranium; an atomic pile for plutonium production; and a heavy water plant. President Roosevelt signed off with a simple, initialed approval to proceed.

The Army Corps of Engineers, which had experience with large construction projects, was to build and operate the factories to create the bomb; the OSRD would manage scientific research and pilot plant development. Placing the project within the Army was a calculated move to help ensure secrecy by hiding it within the Armed Forces and burying its funding within the Corps' large war-time budget. The next day, June 18, Colonel James C. Marshall was given orders to create a new district in the Army Corps of Engineers to carry out the special work. The Manhattan Engineer District, officially designated on August 13, 1942, would oversee the project which was known as the Development of Substitute Materials (DSM).

LABORATORY FOR THE DEVELOPMENT OF SUBSTITUTE MATERIALS

The bomb project was initially headquartered in the offices of the North Atlantic Division of the Army Corps of Engineers on the 18th floor of the Arthur Levitt State Office Building, 270 Broadway in New York City. In August 1942, in an attempt to draw less attention and curiosity to the work, the project's name was simplified from DSM to the Manhattan Engineer District (MED).[3] Colonel Leslie Groves, who had overseen construction of the Pentagon and hundreds of other military projects, was named director of the project on September 17.[4] He began work immediately, relocating MED headquarters from New York City to Washington, DC.[5] Lieutenant Colonel Kenneth D.

5.1. General Leslie R. Groves. © Copyright 2011 Los Alamos National Security, LLC. All rights reserved.

Nichols was named deputy district engineer of the Manhattan Engineer District. Later promoted to district engineer, he was to be responsible for the work at Clinton Engineer Works and Hanford, as well as procurement of materials.[6]

Materials acquisition began quickly. On September 18, Groves ordered the purchase of 1,250 tons of Belgian Congo uranium ore

CIC-9: 85-1790

5.2. J. Robert Oppenheimer. © Copyright 2011 Los Alamos National Security, LLC. All rights reserved.

from the African Metals Corporation. The ore had been shipped to the United States by Belgium's Union Minière mining company to protect it from the Nazis and was stored in 2,007 steel drums in a warehouse on Staten Island.[7] The African uranium comprised about two-thirds of the supply, with additional ore supplemented by lesser quantities from Canada and the Colorado Plateau. The next day, Groves ordered the purchase of 52,000 acres of land in Tennessee for planned uranium separation and plutonium production plants to be known as Clinton Engineer Works.[8] The site was to house facilities for several processes then in the experimental phase. Schedules called for construction of a nuclear reactor to begin by October 1; an electromagnetic plant by November 1; a centrifuge plant by January 1, 1943; and a gaseous diffusion plant by March 1, 1943.[9]

Theoretical research on the materials needed for fission was ongoing at dozens of sites across the country—in private research institutes and in academic and industrial laboratories. The plutonium-239 that would fission to power the bomb was to be produced by irradiating uranium-235 in a reactor.[10] Chemists at the University of Chicago's Met Lab were focused on this process, preparing plans for the large-scale production of plutonium and for its use in bombs. They worked to find a system using uranium in which a chain reaction would occur, determine how to separate plutonium chemically from the other fissionable material, and obtain theoretical and experimental data for an explosive chain reaction with either uranium-235 or plutonium.[11]

In order to safeguard Manhattan Project research, Groves insisted on military compartmentalization, which required that information about the bomb project be separated, much like the individually sealed compartments in a man-of-war ship that protected the whole vessel from sinking when one portion was torpedoed. Workers were to know only what they needed to complete their jobs and nothing beyond that. There was to be little exchange of information. Prohibition against conferring on research was a challenge to scientists, who were accustomed to sharing their findings. To overcome this challenge, a remote central laboratory was needed for weapons physics research and design; a place where scientists could freely discuss their work. Physicist J. Robert Oppenheimer, who had built a school of theoretical physics at the University of California, Berkeley, was brought aboard as scientific director in November 1942. He had considered locations for such a laboratory at sites ranging from Cincinnati to Clinton. In November, a boarding school on a remote mesa in Los Alamos, New Mexico, was selected as the central laboratory site. Construction of the laboratories, which would be known as Site Y, began that month, with staff arriving in March 1943 to barely completed facilities.

INDUSTRY JOINS THE BOMB PROJECT

During the fall of 1942, anticipating the need for industrial-scale production, Groves moved fissile material production out of the hands of scientists and placed it under the management of industry, which could operate on a large scale. Discussion on this matter had been ongoing among scientists from the start of the uranium research project. As reported by chemist Glenn T. Seaborg in his journal, Fermi and others at the University of Chicago's Metallurgical Laboratory had debated whether research would be best handled by scientists, private industry, or by the government; opinions were divided.[12] Most scientists felt a proud ownership of their work and were reluctant to let industrial engineers take over. With the creation of the Manhattan Engineer District and control of the bomb project now in the hands of the Army, however, the scientists would not get to choose. Instead, their input would inform that of the industrial engineers brought aboard to carry out production. E. I. du Pont de Nemours and Company (DuPont) was contracted to construct and operate the pilot reactor in Tennessee that would separate plutonium from uranium based on processes defined by Fermi. Seaborg and other scientists from Chicago's Met Lab would consult on the plant. Industry, which had been charged with profiteering during World War I, was cautious about its engagement in managing the chemical processing facilities. Both DuPont and later Monsanto struck deals with the government to offer services for the sum of one dollar over actual costs. In addition, DuPont vowed to stay out of the bomb business after the war and, like Monsanto, offered all wartime patents to the U.S. government.[13]

Clinton plant design began practically before process development was complete, as Compton led ongoing theoretical work at the Met Lab where Fermi focused on plutonium and fission piles in a quest to understand the physics of the atomic bomb. A milestone was achieved in the research on December 2, 1942, when the first controlled nuclear reaction in history took place in Fermi's Chicago Pile-1 Reactor (CP-1), located in a double's squash court under the west stands of the University of Chicago's Stagg football stadium.

In January 1943, Groves selected a site near Richland, Washington, for full-scale plutonium production reactors. The remote site 130 miles southwest of Spokane would be the Hanford Engineer Works,

5.3. Enrico Fermi, c. 1950. National Archives (Atlanta), records of the Atomic Energy Commission. No. 595043.

or Site W. The reactors and chemical separation plants there—based on Fermi's design and Seaborg's chemical separation processes—would be overseen by area engineer Franklin Matthias and managed by DuPont's Roger Williams. Fermi and Seaborg spent much of the Project shuttling back and forth between Chicago and the production sites.

5.4. Chicago Pile-1 (CP-1). A drawing by Melvin Miller depicts the graphite reactor (Chicago Pile-1) where the world's first controlled self-sustaining nuclear chain reaction took place on December 2, 1942, under the guidance of physicist Enrico Fermi. The reactor was located in a double's squash court under the west stands of the University of Chicago's Stagg Field football stadium. Special Collections Research Center, University of Chicago Library.

With the beginnings of an industrial-scale project in place, President Roosevelt signed off on bomb development on December 28, establishing what ultimately became a government investment in excess of $2 billion. As project leaders were still unsure of the most effective way to create a bomb, the Manhattan Engineer District was authorized to build a full-scale gaseous diffusion (K-25) plant to produce enriched uranium; a plutonium reactor (X-10 was experimental; Hanford was full-scale); and an electromagnetic plant (Y-12), as well as an alternate heavy water production facility. DuPont constructed three heavy water plants to produce neutron moderators. They were located in Morgantown, West Virginia; at the Wabash Ordnance Works in Indiana; and at the Alabama Ordnance Works. Though the heavy water facilities operated, they never produced enough to justify the expense.

Uranium was widely available but required enrichment for use in a bomb. The U-235 isotope of uranium needed for the bomb is present at only 0.7% in natural uranium; the other 99.3% is non-fissionable U-238. The concentration of U-235 had to be increased to around 90% for bomb material, a task given to the K-25 and Y-12 units that would produce the highly enriched uranium-235.[14] Industrial contractors handled the

enriched uranium production at Clinton: K-25 was designed and operated by a specially created division of M. W. Kellogg, Kellex Corporation; Union Carbide took over operations in March 1945. Y-12, which
was based on Ernest Lawrence's electromagnetic studies at Berkeley,
was operated by the Eastman Kodak subsidiary Tennessee Eastman.

The route to plutonium was more complicated than that to
enriched uranium. Plutonium had first been isolated from uranium
on December 14, 1940, by Glenn T. Seaborg, Joseph W. Kennedy, and
Arthur Wahl by deuteron bombardment of uranium-238 in the 60-inch
experimental cyclotron at the University of California, Berkeley. It had
never been produced in the quantities needed for a bomb. Manhattan Project scientists needed to research and develop a method for
producing plutonium in a uranium pile, and then design a way to
separate it in usable quantities. Their work would inform design and
processes at Clinton's X-10 graphite reactor, which would use neutrons emitted in the fission of uranium-235 to convert uranium-238
into small amounts of plutonium-239. The plutonium would then be
separated from the unreacted uranium and fission products in a semiworks chemical separation plant.[15]

Using designs based on Fermi's Chicago Pile, DuPont began
construction of the X-10 reactor on February 2, 1943, with operation
beginning 10 months later on November 4.[16]

On December 20, the first batch of irradiated uranium slugs was
sent to the Clinton chemical separation plant. Plutonium separation
was done using two procedures developed by Seaborg: the bismuth
phosphate method and the lanthanum fluoride method. Both procedures used an aqueous solution to separate plutonium from irradiated uranium; one solution contained bismuth phosphate, and the
alternative used lanthanum fluoride. DuPont ultimately chose the bismuth phosphate separation method for Hanford.[17]

As the plants were built and the Los Alamos research facilities
constructed, administrators organized the bomb effort. In November
1942, Groves appointed a final review committee led by MIT chemical
engineer Warren K. Lewis to evaluate the Project's ability to manufacture uranium and plutonium. The chemistry of plutonium was then in
the hands of the scientists who had discovered it: Kennedy and Segrè
were at Los Alamos, and Seaborg was by that time at the Met Lab,

5.5. L. P. Jernigan loads aluminum-jacketed uranium slugs into a fuel channel on the loading face of the experimental X-10 graphite reactor at Clinton Engineer Works in Tennessee, c. 1947. The reactor had 1,248 channels for uranium slugs. Courtesy of Oak Ridge National Laboratory, U.S. Department of Energy. Photo by Ed Westcott.

where his team focused on processes for the production and separation of plutonium in the piles at Oak Ridge and Hanford. In February 1943, Arthur Compton, head of the Met Lab, summarized the state of chemistry research, noting that purification and production studies of the X-10 "product" (plutonium) would take place at the Met Lab. In the meantime, a chemical lab for experiments on purification and production of plutonium would be built at Los Alamos.[18] Due to the urgency of the bomb program, though, there was no time for the traditional detailed steps used to plan a factory. Construction of the full-scale plutonium reactor had to begin before complete results from the experimental X-10 reactor were available. The Hanford design was based, then, primarily on laboratory-scale information.

CHEMISTRY IS ORGANIZED AND AN INDUSTRIAL LEADER SOUGHT

In the spring of 1943, Manhattan Project chemistry was further refined and organized. In a report issued on May 10, the Lewis Committee advised on Los Alamos research plans and schedules and on organization of the chemistry and metallurgy programs then scattered around the country. The committee recommended that Los Alamos take responsibility for the purification of the plutonium to be used in the bomb and work closely with metallurgists to prepare materials for fabrication. This greatly expanded the role of chemistry in the project and required a reorganization of the Los Alamos work into four divisions.[19] The Los Alamos chemistry staff was to grow from the initially planned six or so chemists to a roster of 30 or more chemists and technicians centralized in a new Chemistry and Metallurgy (CM) Division.

At its peak in 1945, Site Y's CM Division would employ about 400 staff members and technicians.[20] Within the division, a group known as CM-5 was assigned to work on uranium and plutonium purification, while CM-8 undertook uranium and plutonium metallurgy. The purification work called for the design and construction of a new chemistry laboratory—Building D—that would be dust-free and air conditioned to accommodate the sensitive work.[21] Los Alamos, originally conceived as a small community of research scientists, was developing into a large and complex industrial laboratory.[22] And like an industrial laboratory, Los Alamos—and the totality of Manhattan Project chemistry operations—needed a professional science administrator.

Oppenheimer noted the need for coordination between the laboratories and an improved "systematic, discreet and effective interchange of information between this laboratory (Los Alamos) and the other parts of the project." Kennedy was overseeing chemistry at Los Alamos, but Oppenheimer suggested that the lack of overall coordination had led to a situation in which "things of vital interest to us" were not being reported to Los Alamos.[23] "I believe the appointment of a man charged with this responsibility and in whom the higher direction of the project had complete confidence, would greatly facilitate the achievement of this interchange," he wrote.[24] Communication between sites remained an issue throughout the project, even as late as April 1945, as evidenced by a letter from Richard Dodson of the chemistry and metallurgy group at Site Y to Oppenheimer asking that Dayton be permitted to share its semi-monthly technical reports with scientists in Los Alamos.[25]

Partially because of Kennedy's youth—he was 26—and largely because of the project's similarities to an industrial-scale project, Groves and Conant looked beyond Los Alamos for someone with experience in both science and industry to coordinate the chemistry and metallurgy then spread between Los Alamos, Chicago, Berkeley, and Ames.[26] Industrial chemist Charles Allen Thomas, 43, fit the bill for the position of science administrator.[27] He was known by Groves's science advisors Conant and Compton and through his work as a consultant to OSRD director Vannevar Bush. He was also recognized for his work with the NDRC and Rubber Survey Committee.[28] He was cited for his experience directing research at Monsanto, and proven leadership skills.[29] Perhaps most importantly, with his experience in industry, Thomas could arrange for the large-scale polonium purification that would be necessary if polonium were chosen as an initiator for the implosion method then under consideration for the bombs.[30]

Thomas was said to "combine to a greater degree than perhaps anyone in the nation, a brilliant scientific mind and truly extraordinary executive and administrative talents."[31] He was also known for having good people skills, and could bridge the gap between the project director and the scientists, many of whom felt uneasy with the military regulations and secrecy that were alien to the cooperative spirit of scientific investigation and a hindrance to research.[32] He was also said to have a tremendous ability to resolve petty jealousies and to coalesce scientists working under him into an efficient research team.[33] A newspaper offered this description: "He was brilliant in his way of bringing order and understanding in a working, dynamic group, Thomas's friends say. . . . One reason he was so successful was that he was affable, gregarious, considerate and at times just a lot of fun."[34] Unlike General Groves, who had a reputation as a tactless martinet with no respect for science, Thomas was said to be a chemical statesman attuned to successful corporate organization: "His easy chuckle, overworked briar, and air of informality hide a headful of chemical formulas, evangelism, and profound knowledge of power—atomic power."[35]

THOMAS COMES ABOARD

In his role as NDRC chairman, Conant had been in constant communication with Thomas regarding Monsanto's defense contracts. In one exchange, Thomas requested that Monsanto's war projects be given an increased War Production Board priority from AA-2 to AA-1, which was reserved for essential weapons and equipment; only AAA was higher. The Manhattan Engineer District, by comparison, was initially given an AA-3 priority. The request is an indication either of an inflated ego on the part of the industrial chemist or of the increasingly important role the chemical company played in national defense. Conant reminded him that most NDRC projects had AA-2 ratings and some had AA-3.[36] Among Monsanto's NDRC projects in 1943 were:[37]

- Spun-bonded (nonwoven) nylon fabric simulating a knitted fabric introduced in experimental quantities in 1944. The material, which combined high heat distortion with moldability, good strength, good electrical properties, and dimensional stability, was used for battery cases, coil forms, condenser housings, radio crystal holders, radar insulation, and as a replacement for mica-filled phenolics.
- Copolymer developed in collaboration with General Electric that was used by the U. S. Navy Bureau of Ships for radar insulation in the Pacific Theatre.
- Liquid plastic used as a sealant for impregnating magnesium castings to prevent gasoline leakage on aircraft engines.
- Santomerse, an all-purpose and seawater detergent and personal soap for use by the Army and Navy.
- Resproof, W. R., an easily applied, permanent water repellant compound (dichlor ethylbenzene) that gave plastic a high melting point and was of particular interest to the Navy.
- Substitutes for natural waxes, which were cut off during the war.
- A new process for making cotton fabric water repellant.[38]

Early in May 1943, Thomas met Conant and Richard Tolman on the East Coast to witness the test of a new underwater explosive. Shortly afterward, he received a call from General Groves and was asked to travel to Washington, DC, for a meeting scheduled on

May 24 at 11 a.m. When he arrived in the general's office, he found Conant also in attendance.

> After swearing him to secrecy, they revealed the plan to build an atomic bomb. He and Conant spent the day discussing the technical aspects of the question and the probabilities of success. . . . It appeared that the amount of chemistry involved in the project had been under-estimated. They urged him to become co-director of Los Alamos with Oppenheimer and to be responsible for the chemistry of the entire project. To aid him in reaching a decision, a two-day conference was set up at Los Alamos between Conant, Groves, Oppenheimer and himself.[39]

Thomas declined the associate director position, presumably because he did not want to move to Los Alamos; to do so would have meant relocating his wife and children (then ages 6 to 16) or leaving them behind, and it would have required a leave of absence from Monsanto, where he had an established career and was involved with other wartime projects. He was asked again later in the project to move to Los Alamos but once again declined.[40] Instead, he agreed to remain in Dayton and coordinate Manhattan Project chemistry.[41]

> "The coordination of these great research teams was no easy matter. Naturally there were divergent views and healthy debates on which way the research should be directed but Charlie is a great listener and an adroit questioner, and he can present all views of a situation very deftly and clearly. As a result of this work, more is known about the chemistry and metallurgy of plutonium than about many of the elements discovered over 50 years ago," Dayton Project assistant director Carroll (Ted) Hochwalt later observed.[42]

Thomas spent May 31 to June 4 in Los Alamos, meeting with Oppenheimer and Kennedy to further discuss the bomb project and give input on the design and construction of the new chemistry building that would be ready for occupation by the end of the year. Though active in the project, Thomas did not begin his official Manhattan Project duties until July, in part because NDRC policy

5.6. Manhattan Project security badge. The color and letter on the badges indicated the type of clearance; Thomas had top clearance. Charles Allen Thomas Papers, Washington University Libraries, Department of Special Collections.

5.7. Manhattan Project staff button. Charles Allen Thomas Papers, Washington University Libraries, Department of Special Collections.

prohibited Oppenheimer and Manhattan Project leaders from recruiting anyone from its other projects. Thomas visited Met Lab chemistry section head Glenn Seaborg in Chicago on June 25. And a few weeks later, he returned to Los Alamos, this time taking Hochwalt with him. On July 23, Thomas resigned from his NDRC position and promptly assumed his Manhattan Project duties.[43] Conant, the NDRC director to whom Thomas tendered his resignation, was fully aware of why Thomas resigned but was cryptic in responding to the official notice. "I am familiar with the reasons which prompt your resignation. . . . I realize that other parts of the war effort will be the gainer," Conant wrote.[44]

Thomas ended his visit to Site Y—and perhaps celebrated his new assignment—with cocktails and dinner in Santa Fe. He and Hochwalt joined Robert and Kitty Oppenheimer, Seaborg and his wife Helen, and Joe and Adrienne Kennedy for cocktails at the La Fonda Hotel in Santa Fe, followed by dinner at a local restaurant.[45]

CHAPTER 6

PLUTONIUM AND POLONIUM

DURING THE FIRST SIX MONTHS of the program, experimental work at Site Y was focused on designs for two gun-type atomic weapons. One would use a gun system to fire a uranium "bullet" into a uranium "target," while the other would use plutonium bullets and targets. Enriched uranium was to be produced at Clinton, but it wouldn't be until the summer of 1945 that enough was available for one bomb. Plutonium, which had first been synthetically produced in microscopic quantity at Berkeley in the winter of 1940–41, was to be the material for the second bomb. Finding a way to mass-produce plutonium without the impurities from chemical processing that posed pre-detonation risks was one of the key challenges facing Manhattan Project chemists.

PLUTONIUM

Plutonium does not occur in nature but is artificially created by irradiating uranium-238. This process would take place first in the experimental X-10 reactor at Clinton and then on a full industrial scale in

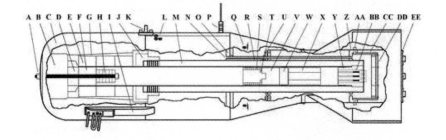

Cross-section drawing of Y-1852 *Little Boy* showing major mechanical component placement. Not shown are the APS-13 radar units, clock box with pullout wires, baro switches and tubing, batteries, and electrical wiring. Numbers in ( ) indicate quantity of identical components. Drawing is shown to scale. ©2016 John Coster-Mullen.

A) Front nose elastic locknut attached to 1.0" diameter cadmium-plated draw bolt
B) 15.0" diameter forged steel nose nut w/14" diameter back end
C) 28.0" diameter forged steel target case
D) Impact absorbing anvil surrounded by cavity ring
E) 13" diameter 3-piece WC tamper liner assembly w/6.5" bore
F) 6.5" diameter WC tamper insert base
G) 18" long K-46 steel WC tamper liner sleeve
H) 4" diameter U-235 target insert discs (6)
I) Yagi antenna assemblies (4)
J) Target-case to gun-tube adapter with four vents slots and 6.5" hole
K) Lift lug
L) Safing/arming plugs (3)
M) 6.5" bore gun tube
N) 0.75" diameter armored tubes containing primer wiring (3)
O) 27.25" diameter bulkhead plate
P) Electrical plugs (3)
Q) Baro ports (8)
R) 1.0" diameter rear alignment rods (3)
S) 6.25" diameter U-235 projectile rings (9)
T) Polonium-Beryllium initiators (4)
U) Tail tube forward plate
V) Projectile WC filler plug
W) Projectile steel back
X) 2-pound WM slotted-tube Cordite powder bags (4)
Y) Gun breech with removable inner breech plug and stationary outer bushing
Z) Tail tube aft plate
   AA) 2.25" long 5/8-18 socket-head tail tube bolts (4)
BB) Mark 15 Mod 1 electric gun primers w/AN-3102-20AN receptacles (3)
CC) 15" diameter armored inner tail tube
DD) Inner armor plate bolted to 15" diameter armored tube
EE) Rear plate (w/smoke puff tubes) bolted to 17" diameter tail tube

6.1. "Little Boy." © Copyright 2016 John Coster-Mullen.

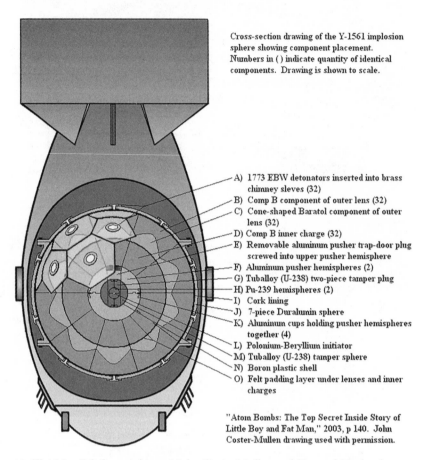

Cross-section drawing of the Y-1561 implosion sphere showing component placement. Numbers in ( ) indicate quantity of identical components. Drawing is shown to scale.

A) 1773 EBW detonators inserted into brass chimney sleves (32)
B) Comp B component of outer lens (32)
C) Cone-shaped Baratol component of outer lens (32)
D) Comp B inner charge (32)
E) Removable aluminum pusher trap-door plug screwed into upper pusher hemisphere
F) Aluminum pusher hemispheres (2)
G) Tuballoy (U-238) two-piece tamper plug
H) Pu-239 hemispheres (2)
I) Cork lining
J) 7-piece Duralumin sphere
K) Aluminum cups holding pusher hemispheres together (4)
L) Polonium-Beryllium initiator
M) Tuballoy (U-238) tamper sphere
N) Boron plastic shell
O) Felt padding layer under lenses and inner charges

"Atom Bombs: The Top Secret Inside Story of Little Boy and Fat Man," 2003, p 140. John Coster-Mullen drawing used with permission.

6.2. "Fat Man." © Copyright 2016 John Coster-Mullen and Howard Morland.

reactors at Hanford. In 1942, all of the world's plutonium could have been piled on a pinhead. The 500 micrograms produced in Berkeley had only been measured, never seen. Because the material was in such miniscule supply, much of the early Manhattan Project work on plutonium was theoretical or done using uranium as a stand-in. "Although we were fully aware of the impurity problem, we conducted little research on it during the first eight or nine months because adequate supplies of plutonium were nonexistent during that period," recalled Los Alamos chemist Ed Hammel.[1] The rare element was in high demand. In June 1943, chemist Glenn Seaborg noted that a request made to the Met Lab by Site Y for 200 micrograms of plutonium was a hardship. Of its existing supply, Chicago had used 30 micrograms for extraction process studies for Hanford, 50 micrograms had been sent to Berkeley, and planned experiments at the Met Lab would require 10 micrograms each.[2] It was not until January 1944, after the Clinton X-10 reactor began producing plutonium, that scientists had enough material for experimental use. Even then, the plutonium was "remelted, repurified and re-reduced many times as we tried out, on gradually increasing scales, the processes that had been developed with 1-gm lots," wrote Los Alamos metallurgist Cyril Stanley Smith.[3] The uranium and plutonium products were valuable, worth an estimated $10,000 per gram, and a little product went a long way.[4] By August 1944, Site Y had received 51 grams of plutonium from X-10, which was used for approximately 2,500 experiments.[5] Among Thomas's many roles as coordinator of the Project's chemistry and metallurgy was mediating the competing requests from scientists at the various sites for the scarce amounts of plutonium. "I realize there are many demands on the meager stock of plutonium now on hand in Chicago," he stated.[6]

Despite a relative lack of knowledge about plutonium, the scientists were aware from the start of the project of a fundamental problem—plutonium was likely too unstable for the gun design and could create a low-efficiency explosion that would threaten to pre-detonate before the projectile and the target met to activate the bomb. Seaborg had cautioned as early as November 1942 that the plutonium fission process could produce impurities capable of setting off a pre-detonation, and Conant had reported this to the OSRD's S-1 Executive Committee.[7]

POLONIUM

In looking for ways to avoid pre-detonation, Site Y physicist Seth Neddermeyer had proposed in April 1943 that instead of relying on a gun assembly, the chain reaction inside the atomic bomb could be more predictably triggered by implosion, or inward explosion. His idea, which was uncharted territory, was not supported, and work moved forward on the gun assembly. In the meantime, Neddermeyer pursued implosion.

In the implosion method, an initiator at the center of the fissile core would produce a rapid burst of neutrons to initiate a fission chain reaction at the optimal time.[8] The best-known neutron sources were radium-beryllium and polonium-beryllium. Radium and polonium emit alpha particles that strike beryllium to produce neutrons and initiate the explosive chain reaction. In his introductory lectures to incoming Los Alamos scientists, physicist Robert Serber had suggested using radium-beryllium in a gun bomb with the radium attached to one piece of core material and the beryllium to the other; they would be arranged to smash together when the gun was fired, and the two core components mated to complete a critical assembly.[9] For the implosion-model plutonium bomb, an initiator, or "urchin," made of polonium and beryllium separated by layers of nickel and gold would be at the core of the bomb and would serve as a neutron source to initiate a chain reaction. Conventional explosives would compress the plutonium core surrounding the triggering device and break the nickel/gold barrier between the polonium and beryllium. This would allow alpha particles from the polonium to strike the beryllium, producing the neutrons to initiate a nuclear chain reaction.[10] Polonium was optimal for a trigger material, because its rate of alpha emission is about 4,000 times greater than that of radium. By the end of 1943, the implosion method had gained enough support to put it on equal standing with the gun-assembly design in priority.[11]

In June 1943, Oppenheimer wrote to Groves, defending the need for polonium: "We shall make an effort in detonating our bomb to reduce the neutron background to avoid pre-detonation. We have, therefore, artificially to produce a burst of neutrons at the moment when the position of the materials involved is optimal. . . . A source of this strength involves several curies of polonium in adequate contact with beryllium."[12] He explained that polonium could be retrieved in two ways: from lead residues in which it occurs naturally or artificially produced from bismuth.[13] Polonium had not to that time been isolated in pure form or appreciable quantity, but Oppenheimer indicated that Wendell Latimer in California would pursue how to extract polonium from bismuth irradiated in the UC Berkeley cyclotron.[14]

Polonium-210 (Po-210), a strong emitter of alpha particles, was discovered by physicist Marie Curie in 1898. It occurs naturally in sources such as lead-containing wastes from the uranium-238 decay

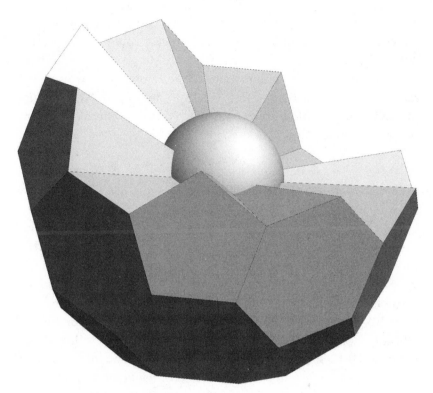

6.3. Initiator. © 2016 Copyright John Coster-Mullen.

series, vanadium, and radium refining operations. With a 138.39-day half-life, it was intense enough to be useful for the Project but not long-lived enough to be stockpiled. In addition to being retrieved from radium in lead dioxide residues, polonium could also be produced by bombarding bismuth with neutrons. The conversion of bismuth-209 to polonium had first been reported in 1935. Bismuth-209 irradiated with neutrons forms bismuth-210, which decays rapidly into polonium-210. This process, though unexplored, seemed to be a more promising supply of polonium for the bomb project than the tedious and material-intensive method of extracting polonium from naturally occurring lead ores.[15]

Polonium-210 was to be produced by bombarding bismuth in Clinton's X-10 reactor on an experimental basis, with the first samples due by January 1, 1944.[16] Once the process had been worked out on an experimental scale, it would be produced on an industrial scale at Hanford. In the meantime, to ensure a supply of polonium in the event that the bismuth phosphate process could not be worked out, Compton suggested that Thomas and Monsanto move forward with production of polonium from the lead ores. It was estimated that Latimer would require four weeks to develop the lead dioxide method, which would then be implemented by Monsanto, where processing would take four weeks.[17]

BIRTH OF THE DAYTON PROJECT

On May 24, 1943, contract W7407eng18 with Monsanto Chemical Company became effective, covering the Manhattan Engineer District's research and development of polonium. The contract, administered by the Chicago Area Engineer of the Manhattan Engineer District, effectively created the Dayton Project. While Monsanto President Edgar Queeny agreed to the contract, the corporation did not officially acknowledge the collaboration—nor did the Manhattan Engineer District, due to the extreme secrecy of the work. The work in Dayton was so essential to the success of the bomb project and so deeply hidden within the already-secure Manhattan Project that Dayton researchers were not recognized as part of the greater project and could not obtain materials through usual Manhattan Engineer District networks.

In late July, Groves outlined Dayton's responsibilities in a letter to Thomas:[18]

1. The purification of the final product. By purification I mean the purification from approximately 99% upwards.
2. The production of polonium; it being understood that the radiation of the bismuth will not be under your charge.
3. The preparation of the hydride, in the event that such preparation becomes necessary.[19] In each case the research is included as well as the actual production.[20]

With orders in hand, Thomas established monthly meetings on plutonium and polonium research in Chicago that were attended by representatives of each site; scheduled monthly visits to Los Alamos;

and had weekly consultations with Groves. He was aided in Chicago by his deputy, John C. Warner; and by Joseph Kennedy at the Los Alamos lab. In August 1943, he began sending monthly "progress reports" to Groves and Conant, updating them on plutonium and polonium activity at the various sites under his control. These monthly reports, which continued through July 1944, provide a valuable summary of the chemistry of the Manhattan Project.

CHAPTER 7

THE DAYTON PROJECT COMES TO LIFE

BY LATE SUMMER 1943, the Dayton Project was up and running. Monsanto Central Research at 1515 Nicholas Road organized polonium production. Room 30 served as the center of operations and was used for interviews, conferences, supply ordering, a research library, and small-scale laboratory work.[1] Thomas was project director; his longtime business partner, Carroll A. Hochwalt, was named assistant project director; and James H. Lum of Monsanto, who had worked in the NDRC explosives research laboratory, was appointed laboratory director and chief of staff. Nicholas N. T. Samaras, who had a doctorate in chemical engineering from Yale and had overseen Monsanto's styrene research, also joined the leadership team. W. Conard Fernelius, a Purdue University chemistry professor, was brought aboard as assistant laboratory director.

The staff got to work immediately. Hochwalt met in Los Alamos with Oppenheimer, Kennedy, Segrè, and chemist Arthur Wahl from July 20 to 25, 1943, to discuss the ways in which Dayton could ensure an adequate polonium supply.[2] Two methods were identified as best for meeting the required production quotas: polonium could

7.1. Charles Allen Thomas, left, and Carroll (Ted) Hochwalt. Courtesy of the Thomas family.

be synthesized by bombarding bismuth with neutrons, which to that time had only been done experimentally at the Berkeley cyclotron. It could also be extracted from naturally occurring plattnerite, a lead-containing mineral; polonium was produced from the decay of radioactive lead. The extraction process involved first removing the lead in the form of lead dioxide from the mineral and then extracting polonium from the lead dioxide. The lead dioxide process would be explored in Dayton while scientists in Berkeley completed their experimental work on the bismuth process. The bismuth process would be implemented first in Clinton's experimental X-10 pile and then on an industrial scale at Hanford. Oppenheimer initially estimated that 10 pounds of bismuth irradiated in the X-10 pile would yield 1 curie of polonium, and that 100 pounds would yield 9 curies every four months, if the pile were operated at 20KW/ton of fuel.[3] While that process was developed, Dayton explored the lead dioxide process using lead dioxide residues from the Port Hope radium refinery in Ontario owned by Eldorado Mining and Refining

7.2. James H. Lum, left, and W. Conard Fernelius. *Dayton Daily News*, August 7, 1945.

Limited, a Canadian government enterprise.[4] The lead dioxide came from African and Canadian uranium ores, which were estimated to contain about 1 curie of polonium per gram of radium extracted from the residues.[5] It was later estimated that 6 tons of Port Hope lead dioxide might produce 0.2 or 0.3 mg of polonium per ton.[6] In a letter to Groves, written July 27, Oppenheimer stressed the importance of acquiring the lead dioxide reserves before they were taken by petroleum companies, which used lead dioxide to prospect new wells. He also concurred with the decision to have Dayton proceed with the work.

"It would seem to be in accordance with the general decision to give Dr. Thomas overall responsibility for the polonium production and extraction that this matter be put in his hands, and that such investigations as seem necessary on this extraction from the lead residues be done under his direction," Oppenheimer wrote.[7] In a letter to Groves a few weeks later, Thomas requested that the lead dioxide be held in Canada for three months, while the bismuth process was further explored.[8]

Thomas went to the Met Lab in Chicago to review chemical and metallurgical programs, and reported to Oppenheimer on July 31 that he was eager to hear about the status of Latimer's work with bismuth in Berkeley. "If it is necessary, we can go into this matter with Latimer to see that there are no loose ends and that this work will be completed with dispatch," he wrote.[9] By early August, Oppenheimer, Groves, and Thomas were conferring on the progress of the polonium research—or lack thereof—in Berkeley. In a letter to Thomas written August 5, Oppenheimer expressed frustration with the pace of work in California. "I am sending you a copy of a letter to Latimer from which you will see that we ourselves do not know too much of what is going on there. Since the problem of polonium extraction from bismuth does not lie in the field of Latimer's main interest, and since he is rather short-handed, I think that it may require a little pressure from us to get this work done in time to be of use."[10] Hochwalt departed for Berkeley on August 15 to check in on Latimer, stopping in Los Alamos on his way home. "I think it would be a good thing for him to bring you up to date on our activities which have been rather hectic and I find them difficult to describe in detail by correspondence," Thomas wrote to Oppenheimer about Hochwalt's trip.[11]

Communication in August 1943 from Thomas to Groves and Conant included input on staffing issues and the effort to get enough scientists on board at all sites to do the proposed chemistry. Latimer, it was noted, had eight chemists in Berkeley working on plutonium and was placing two on the extraction of polonium from bismuth; Dayton would send an additional two men to California to aid in the effort and learn about the process.[12] This call for a staff increase required shifting chemists from one site to another and freeing Kennedy from his role supervising the Los Alamos service chemistry group so that he could lead purification efforts in the new Chemistry and Metallurgy Division. "Mr. Oppenheimer and I feel that a good man is needed to head up the chemical and metallurgical groups at Y," Thomas wrote to Groves.[13] He also noted that the Los Alamos laboratory was too small for the new assignment: "These men will be very crowded but by rearranging and conserving the present space, the work can be handled even though the service group has been increased," he wrote.[14] Accordingly, the new D laboratory was planned, and Lum and Samaras were sent from Dayton to Los Alamos to lend a hand and get a better understanding of the work. Lum couldn't tell his wife where he was going; he could only leave her an emergency phone number. "We thought we were in a race with the Germans to develop the bomb first, but Hitler had decided to spend the money on conventional weapons instead," he recalled.[15] Fernelius worked with Lum to rapidly assemble a team of chemists for the Dayton work; 12 were needed. Fernelius, who had taught at Ohio State University, looked to that university for talent.

RECRUITING SCIENTISTS

Joseph J. Burbage, one of many Ohio State recruits, was the first Dayton Project team member to arrive, reporting to work on August 21, 1943.

Eugene Rembold, an Ohio State chemistry graduate student, was interviewed on a Friday and began work the next Monday, having obtained temporary security clearance. He recalled that nothing was said about the precise nature of the work, but he knew it was secret and sounded interesting.[16] Most Manhattan Project scientists, including those in Dayton, were young. The average age of the researchers in the Site Y chemistry and metallurgy division was 29; many in Dayton were under 25.[17] Finding top scientists took effort, especially when it meant removing them from teaching positions. In September 1943, one of those asked to join the Dayton Project was Louis Marchi, a chemist on the Indiana University faculty who had a doctorate from Ohio State University. Thomas put in a request for Marchi's leave with IU President Herman B Wells that gave little information on the project:

> The work will be done in the field of inorganic chemistry. As you know very few chemists have been trained in this field. . . . I believe it could be of great benefit to Dr. Marchi and the University if he could participate in this project. The work will involve an interesting new development in the field of inorganic chemistry. I am sure that Dr. Marchi would return to Indiana with very valuable experience and that he would have leads for research in his chosen field for many years to come.[18]

Marchi's release ultimately required a letter from Conant, assuring Wells of the importance of the project:[19] "The nature of the research

7.3. Joseph J. Burbage. Special Collections and Archives, University Libraries, Wright State University, Dayton, Ohio.

problem cannot be divulged as it is of a highly secret nature, but you will perhaps be willing to take my assurance that I know of none that has a higher priority in the entire scientific program that is being operated either from OSRD or through the Army or Navy."[20]

Robert A. Staniforth, who had earned a chemistry doctorate at Ohio State University and written a Master of Science thesis in 1942 on nuclear fission and the transuranium elements, was highly sought

7.4. The library at Thomas & Hochwalt Laboratories. Courtesy of the Thomas family.

by the Dayton Project. At the time, he worked for General Aniline & Film Corporation in New York, which produced dyes and film.[21] The company claimed to be involved in sensitive war contract work and was reluctant to lose a top researcher. Staniforth, himself, was reportedly hesitant to depart, unless official word convinced him of the higher priority of his new assignment: "I have no great desire to sever connections with the company or to jeopardize my association in any way. However, I realize that the Dayton Project may be more important at the present time and above personal considerations."[22] Conant convinced the company to release Staniforth, who arrived in Dayton in January 1944.

Between June and December 1943, personnel in Dayton grew from the original estimate of 12 chemists to 46. Also joining the research staff in 1943 were Monsanto's Ross W. Moshier (PhD 1934, Michigan), Carl L. Rollinson (PhD 1939, Illinois) from DuPont in Cleveland, and Edwin M. Larsen (PhD 1942, Ohio State), a chemist on leave from the University of Wisconsin who with Fernelius developed polonium purification processes.[23] By war's end, the Dayton Project would grow to 201 employees.[24] The recruits included chemists as well as physicists and technicians. Practically none of the scientists had prior experience with the new field of radiochemistry or with plutonium and polonium. When they weren't in the laboratory, the scientists could be found in the library, learning as much as they could about nuclear energy.[25]

"Nobody had ever seen (polonium) before. It was a soft, silvery looking metal. If you turned off the lights, you could see a faint purple glow, which would intensify as the purity increased," Fernelius recalled.[26]

A QUICK LEARNING CURVE

The period of October through December 1943 is referred to as "the Conference Period" in the government record of the Dayton Project.[27] It was during this time that much of the Dayton staff's fundamental education about polonium and plutonium took place, either off-site or in meetings with visitors from Los Alamos and other laboratories.[28] Among the major uncertainties in the polonium purification process were how to control the purity of the polonium produced and the efficiency of the various possible chemical separation procedures.[29] Lum spent a week in Chicago learning about plutonium chemistry. Fernelius, who was responsible for the chemical processes of the extraction and purification of polonium, was in Berkeley for a month and returned with techniques and a process. In October, Los Alamos chemist Rene Prestwood visited Dayton to update scientists on polonium purification work.[30] In keeping with the secrecy of the work in Dayton, Oppenheimer's staff travel report to Groves noted Prestwood's trip to Ohio only as "special work."[31] With leadership of the Chemistry and Metallurgy Division in Los Alamos still to be determined, Thomas sent Hochwalt to Los Alamos in October to assist in coordinating the work there.[32]

The scientists and Dayton Project staff worked six days a week and every holiday, except Christmas. Thomas did, too, often adding Sunday to his workweek. He traveled almost constantly, most often by train, but he also had a private pilot's license and owned a four-seat Waco aircraft (purchased in 1938 and christened the Bald Eagle) that he kept at Dayton's Deeds field; he usually left the piloting to a man named "Slim" Camel. From his first visit to New Mexico in May 1943 through the summer of 1944, the third week of each month was spent in Chicago and at Site Y: Monday, Tuesday, and Wednesday were devoted to meetings at the Met Lab, with Tuesday afternoon focused on plutonium and polonium research. Thomas then hopped on a train and spent Thursday through Saturday at Site Y.[33] The Site Y chemistry division leaders Kennedy and Smith often attended the Chicago meetings. "I remember that these monthly 'Thomas group' meetings were held at the Metallurgical Laboratory in Chicago over a year's period extending from the summer of 1943 to the summer of 1944. A great deal of first-class scientific work was carried out in this connection under Charlie's direction,"[34] Seaborg recalled.

KEEPING THE SECRET

Although Thomas had been brought aboard to help coordinate work among the Manhattan Project chemistry laboratories, he quickly learned that this assignment was hindered by Groves' insistence on compartmentalization. Strict security policies governed the exchange of information between sites, but the chemists in Los Alamos and Chicago needed to communicate. Thomas expressed his frustration to Oppenheimer, when writing to ask that Kennedy and Smith from the Los Alamos chemistry and metallurgy division be permitted to attend the monthly Chicago meetings. "It is my understanding that representatives from Y can attend meetings at Chicago, but that representatives from Chicago cannot attend any meetings at Y. Not to have it a reciprocal arrangement may defeat the spirit of cooperation," he wrote.[35]

Permission was given for the exchange of information between specific members of the two sites or by the visit of Los Alamos personnel to Chicago. Information was restricted to chemical, metallurgical, and certain nuclear properties of fissionable and other materials. It was permissible for the representatives to discuss schedules of need for and availability of experimental amounts of uranium-235 and plutonium-239. No information could be exchanged on the design or operation of production piles, the design of weapons, or to permit comparison of schedules of need for and availability of production amounts of active materials. Three members of the Los Alamos Laboratory were to be kept informed of the time estimates for production of large amounts of these materials. In addition to the above, special permission would be granted by General Groves for visits to Chicago by other members of the Los Alamos Laboratory.[36] The rules went all the way to the top and applied in many cases to Oppenheimer as well as his deputies. The need for careful information on time schedules of production was one of the areas most challenged by compartmentalization. Estimates received during the summer of 1943 were vague, incomplete, and contradictory, so it was difficult to make sensible schedules of bomb research and development. This situation smoothed out over time as communication channels were opened at the highest levels.

While letter and Teletype formed much of the communication, written record was generally discouraged for security reasons, and reports often contained incomplete information. Project administrators and section leaders spent much time traveling to discuss research in person. Travel time between Manhattan Project sites was immense and is an often-forgotten portion of the commitment made to the work. Plane travel was rare and limited to the top administrators; train travel was the norm. Travel from Washington, DC, to Oak Ridge took 12 hours; from Chicago to New Mexico was a 24-hour ride. Most of the scientists traveled on the famed (Southwest) "Chief" that ran from Chicago to Los Angeles. They disembarked at Lamy, New Mexico, 10 miles from Santa Fe, and continued by car or bus to Santa Fe or Site Y. Groves was known to take a train from Washington, DC, on an evening, with his secretary riding along for a few stops so he could keep working. She would then get off and return to the Capital on an eastbound train while he continued west.[37] Oppenheimer and Groves had their own train cars. Thomas had a strong dislike of train travel, which he viewed as an inefficient use of time. He was always in a hurry and was known to be absent-minded about travel, reportedly once boarding a train going to Chicago instead of New York and often leaving personal items in hotels and trains, which required his secretary to retrieve them.[38] At the least, he was gone from home two weeks of each month.[39] (See Appendix II for project-related travel.)

Thomas's son, then 16, was sent from Ohio to Deerfield Academy boarding school in Massachusetts. His wife, assisted by household staff, was at home with their three girls, who ranged in age from 6 to 15. In no case did the family know what Thomas was doing. The family assumed Thomas was occupied by Monsanto's artificial rubber work and NDRC contracts. "Until the first atomic explosion, none of us knew that there was a secret to keep. The unclassified Buna-S program (the government rubber work) provided an ideal diverting subject to talk about," his son recalled.[40] Monsanto duties provided excellent cover for the classified work. Newspaper coverage of Monsanto's work at the time includes publicity on the development of a launching propellant for American models of the robot bomb and construction of a sulphuric acid plant to provide for explosives needs.[41] Thomas kept up industry appearances, such as an address to a Harvard Business School Club dinner in October 1944 that was reported by the *Dayton Journal*: "Though synthetic rubber is playing an important part in our war it is still in its infancy as an industry. Wait three years and see what chemistry develops in this line," he told the audience.[42]

Other Dayton Project personnel also traveled frequently. Samaras visited Los Alamos monthly from September through December 1943. Fernelius, the assistant laboratory director whose chief destination was Chicago's Met Lab, did not reveal anything about the nature of his wartime work to his family but recalled that his son was scared nonetheless. "I don't think my wife is completely reconciled to it even yet. And I never realized how much it bothered my son (who was then nine years old). He was very young and had fears about me being the victim of German or Japanese terrorists," he later recalled.[43]

ARMY RECRUITS ADD SCIENTIFIC SKILL

As Dayton's research effort grew, so too did personnel and the scope of the work. This required the addition of men with technical training and scientific education who had been drafted by the Army and then assigned to its Special Engineer Detachment (SED). The detachment had been created to overcome the short supply of skilled workers for wartime projects. By 1945, there were 3,500 or so SEDs across Manhattan Project sites, some 34 of which were assigned to Dayton.[44]

7.5. John J. Sopka. Courtesy of the Sopka family.

John J. Sopka was in his senior year of physics at Harvard when the U.S. Army Corps of Engineers contacted him. It was December 1941, and he was directed to meet with representatives in a room in the basement of Columbia University's Pupin Laboratory of Physics. Sopka's mysterious interviewers told him that he met the qualifications for their needs, but he was given no other information than to wait for further word.

By June 1942, when Sopka graduated from Harvard, he had heard nothing and went to work at MIT's Radiological Laboratory and then to Princeton University, where he taught introductory physics to officers in the Navy V-12 College Training Program and enlisted men in the Army Specialized Training Program. In 1943, still having heard nothing, Sopka received a letter from the draft board ordering him to report for immediate induction into the military. At the same time, he was called to another meeting and told to report within two days to Dayton. The next morning, with little more information, he boarded an overnight train to Ohio.[45] With a background in mathematics, physics, and electronics, Sopka was assigned to the physics staff at Dayton's

Unit III laboratory.[46] He made a quick return trip to Princeton, where he and his pregnant wife packed up and took a train back to Ohio. They settled in Dayton on the upstairs floor of a private home.[47]

Richard Yalman, a Harvard chemistry student and private in the SED, was placed in Dayton in June 1944.[48] Like other SEDs assigned to the laboratories, Yalman lived and worked as a civilian in plain clothes and without military privileges, except when he went to the PX at Patterson Field or was on furlough. This double life could get complicated: a visit to the Army's dental and medical clinic had to be made in civilian clothing; a visit across the street to the PX was made in uniform. If Yalman was doing both in the same day, he had to drive out into the country and change clothing before returning to the base. Betty Halley, who had worked on the Patterson Field flight line servicing airplanes prior to joining the Unit III electronics laboratory, recalled watching the SED costume-changing with amusement: "One of my favorite memories was watching the GIs get ready to go out to Wright Field for whatever it was. They had to change from civilian clothes to their uniforms in van en route."[49]

Yalman recalled an infraction of this dual identity committed by Sam Jones, another young SED chemist on the Dayton Project, who went rowing one day in uniform on a nearby lake. He was picked up and broken from sergeant back to private, with the admonition that he was lucky he wasn't being shipped to the Pacific, Yalman recalled. In another instance, the landlady of a rooming house called the police to report a spy, because she had found an Army uniform hidden away in a coat closet. The man in question had to be transferred to another unit.[50]

Yalman was married in Dayton during the Dayton Project, with Fernelius serving as best man. Yalman and his wife rented a room at 7 Ivanhoe in Oakwood before settling in an apartment on Central Avenue. He hitchhiked from the apartment to the Unit IV facility, preferring that to a 45-minute bus ride. Finding housing was such a challenge for Dayton Project personnel that some scientists left their families behind. Harry Weimer (PhD 1933, Ohio State) was teaching chemistry at Indiana's Manchester College when the war broke out. He transferred to Ball State University in Muncie, Indiana, where he was teaching Army engineers when he was recruited by Fernelius, who had been his OSU doctoral advisor. "Would you participate in a government-sponsored program involving a totally new field of

7.6. Harry Weimer and family, 1940. Courtesy of
Robert Weimer.

science with no questions asked or answers given?" Fernelius asked.
Weimer agreed and joined the Dayton Project in January 1944 to work
in Unit III, where he developed the lead dioxide process. His wife and
sons, ages 11 and 13, remained at home in Indiana.

Weimer had a room in a boarding house but returned home every
third weekend, taking a bus from Dayton to Fort Wayne, Indiana.[51]
"Harry was given weekend leave every three weeks and I received
a gas ration card to pick him up in Fort Wayne. I knew he was under
oath, had daily physical checks, and that his roommate was an Army
colonel who could not wear his uniform publically. They worked
24-hours-a-day in an isolated building. That was about all I knew,"
recalled his wife, Orpha Weimer.[52]

Physicist Sergio De Benedetti and his new bride rented a small,
furnished house at 2716 Athens Avenue in Dayton from a military

family that had relocated for the war. They socialized with a group led by the Ferneliuses, who hosted a monthly gathering to discuss current affairs. De Benedetti, an Italian Jew, had fled Mussolini's Italy and showed up on the doorstep of the Curie Institute in Paris. He had earlier had a fellowship at the institute and was able to rejoin the lab, working in radioactivity. When the Nazis invaded Paris, De Benedetti left the city by bicycle and made his way to the United States via Portugal. He was recruited to the Dayton Project from Kenyon College, where he was teaching Army Air Corps recruits.[53] At one of the Ferneliuses' monthly gatherings, the Italian physicist spoke about his experiences in Italy and the political situation there.[54] His wife, Emma, recalled being brand-new to the United States, recently married, and finding herself in the unfamiliar landscape of Dayton. She was also pregnant with the couple's first child, who was born in Dayton in 1945. She knew nothing about her husband's work.[55]

Other scientists arrived in Dayton with children and came up with creative solutions to the housing shortage, such as that chosen by three families who shared a 16-room mansion in Oakwood. The living room of the home contained a full-size pipe organ, adequate for a large church. The fireplace was large enough for children to hide behind the andirons, and the home came with a four-car garage. None of the recruits had a car, so the wives used children's wagons to carry groceries home from the store.[56] Unit III electronics technician Betty Halley was among the guests at the house, where frequent parties filled the ballroom and picnics were often held on the patio. Conversation at the parties was all about families, with not a word spoken about work, according to machinist Howard DuFour, also a guest at the parties.[57]

BONEBRAKE LABORATORY

Like housing in the greater Dayton area, which was scarce, lab space for the expanding Dayton Project was also at a premium. In order to accommodate the growing work, a lease was signed in September 1943 with the Dayton Board of Education for a 3.5-story brick building at 1601 W. First Street. The building had been constructed in 1879 by the Church of the United Brethren as a home for seminarians attending Union Biblical Seminary.

7.7. Union Biblical Seminary at First and Euclid Streets in Dayton, 1879. The school was renamed Bonebrake Theological Seminary in 1909. U.S. Department of Energy/Office of Classification.

The building was sold to the Dayton Board of Education in the 1920s and served as the Grace A. Greene Normal School. When the Dayton Project took over, the space had last been used as a book warehouse and was in bad shape; the windows were broken, and the staircase between the second and third floors was missing.

Extensive remodeling was required to accommodate the project's chemistry, physics, and electronics laboratories. The facility provided 6,000 square feet of laboratory space on two floors, with an additional 3,000 square feet available on a third floor.[58] Staff initially occupied the first two floors, which were updated with new flooring, re-plastered walls, new windows and sashes, and new systems for heating, lighting, and power. A counting room for measuring radioactivity was added, as were glassblowing and machine shops. Due to the top-secret nature of the work, it was a challenge to obtain lab supplies. To ensure that the work in Dayton was well secured, the site had no official ties to the Manhattan Engineer District. This meant that personnel could not get materials through normal War Department

7.8. The Bonebrake building had last served as a Dayton Board of Education warehouse when it was leased for the Dayton Project in 1943. U.S. Department of Energy/Office of Classification.

7.9. Bonebrake (Unit III), c. 1944. A guardhouse and single-story building that housed the cafeteria are in the foreground. U.S. Department of Energy/Office of Classification.

7.10. Unit III site map, September 1946. U.S. Department of Energy/Office of Classification.

channels, so lab supplies initially consisted of a bushel basket filled with assorted chemical glassware. The situation later improved because of Monsanto's NDRC projects, which provided a good cover for acquisition of materials and equipment. Outside of the building, two guardhouses were built, as was a small chemical storage shed and a security fence. The site, known as Unit III, was near downtown Dayton, in the middle of a working class neighborhood. It eventually grew to include 20 smaller buildings covering nearly a city block.[59]

The first Dayton Project staff members moved into Unit III on October 18, 1943.[60] Unit I, on Nicholas Road, retained administrative offices and X-ray and spectrographic work on polonium through 1949.[61] Robert Staniforth worked on the volatility of polonium in a basement lab at Unit III. Other chemistry labs were located on the second floor,

7.11. Mary Lou Curtis, c. 1948. Courtesy of Bill Curtis.

as was a health physics lab. Monsanto physicist Donald L. Woernley (PhD 1943, Yale) was the first to lead the physics group. Josef W.
Heyd (PhD 1937, Pennsylvania State College) oversaw the electronics
section and was joined in December 1943 by John Sopka, who had a
background in mathematics, physics, and electronics. The two groups
developed and constructed electronic equipment to measure alpha,
beta, gamma, and neutron radiation, built counting and calorimetry
equipment, and constructed instruments for the health group. Health
physics, which later occupied the third floor of Bonebrake, worked
closely with its counterpart at the Met Lab and included the few
women on the technical staff. Among them was radiation physicist
Mary Lou Curtis, who joined the Dayton Project in December 1943.

Curtis was married to a solider who was overseas, and had a
young daughter. She had an undergraduate minor in physics and a
master's degree in mathematics and was looking for employment.
She interviewed with Monsanto for a secretarial job but couldn't
type, so was instead offered a position in the Dayton Project laboratory. "I went into a lab immediately, and this developed into what
we called the 'Counting Room.' We were counting particles or X-rays

that came off of radioactive material, and we had to determine the purity and the amount," she said. As the work in Dayton progressed, her work became more complicated. "As we got different materials to work with or different radioactive elements that we were interested in, I had to develop the techniques for measuring them. Once the techniques were developed and proven, then the girls in the lab did the actual counting work. But I had worked out the procedures first."[62] The Electronics Division designed and manufactured the measuring instruments, because the field was so new that none were commercially available.

POLONIUM EXTRACTION

Thomas's work coordinating the Project's overall plutonium chemistry continued as the Dayton polonium work got underway. The Manhattan Project chemistry division's Project Council met for the first time on Wednesday, October 20, 1943, convening at the University of Chicago in Room 209 of Eckhart Hall. Thomas, who was chair, urged that investigation of the density determination of plutonium take precedence over other phases of the work.[63] The remainder of the week was devoted to meetings focused on separating polonium from lead dioxide. The meetings, which took place October 22 through 25, were attended by Canadian chemist Friedrich Paneth, an expert on separating polonium from lead oxide; Lum and Fernelius from Dayton; Met Lab director Arthur Compton; French-born Bertrand Goldschmidt, who worked with Glenn Seaborg at the Met Lab on separating plutonium and uranium; and Los Alamos physicist Emilio Segrè, who was given special security clearance to attend. [64] In order to keep the polonium work in Dayton under tight cover, General Groves requested that Compton be the one to call the meeting with Paneth and Goldschmidt—rather than Thomas—to give appearances that the Met Lab was the party interested in polonium.[65] A cryptic telegram from Oppenheimer alerted Thomas to Segrè's presence at the meeting. "Arrangements approved for participation by man from here in your conferences with distinguished northern visitor."[66]

Preparations for the start of operations in Dayton continued. On November 1, Fernelius and newly appointed Dayton health physicist Louis B. Silverman conferred with scientists in Chicago on health safety. As Seaborg of the Met Lab observed, health safety measures were basic at best. In May 1944, curative measures included "vacation, transfusions and hope," according to Seaborg. "Early in the stages of the program two hazards were realized, radiation and toxicity. There was a considerable body of knowledge on radiation effects, and the necessary corollary studies were immediately undertaken and have been largely completed. In the case of toxicity, it was indicated that the hazard is serious. This led to a program of toxicity studies which is still continuing," he wrote.[67]

With polonium production set to begin in Dayton, Thomas assured Oppenheimer that staffing was complete; 11 chemists were on hand, and the Dayton "production program" was prepared to furnish 10 curies of polonium per month.[68] He advised Groves that the lead dioxide in Canada should now be purchased as a safeguard against production trouble with the X-10 pile or unexpected difficulties in the recovery of polonium from bombarded bismuth.[69] On November 5, Groves arranged for the purchase and shipment of lead dioxide from Canada's Port Hope refinery.[70] The first shipment—6,250 pounds—arrived in Dayton on November 10. The next day, Benjamin Scott from the Met Lab arrived in Dayton to instruct project members in the setting up, calibration, and use of electronic counters to monitor radioactivity. Rudimentary health and safety measures were instituted, and work began on the lead dioxide process five days later. Oppenheimer had by then increased his polonium request to 20 curies due by September 1944, though he noted that an earlier delivery in July would be preferred.[71] A month after starting production—by December 8—Dayton had processed the first 500-pound batch of lead dioxide, and on December 15, 30 microcuries of polonium were available for electroplating experiments, a miniscule but vital quantity.

Dayton chemist Harry Weimer shared an anecdote many years later about transporting the lead dioxide from the Army Air Corp's Patterson Field to Unit III. He and some SEDs assigned to work on the lead dioxide process "strapped guns about their waists, drove their trucks to Wright-Patterson Airport, and briskly ordered all persons safely back while they loaded on a harmless shipment of raw rock from Canada before a gaping audience."[72] The rock in question was most likely Port Hope lead dioxide residues. Dayton scientists worked simultaneously on the bismuth process. In September 1943, 1,000 pounds of pure bismuth slugs had been extruded by American Smelting and Refining Company and delivered to Dayton, which forwarded 500 pounds on to Clinton for bombardment in the reactor there.[73] The plan was for some 400 pounds of bismuth to be inserted in the X-10 pile monthly and then shipped to Monsanto.[74]

Even with work under way at Unit III, remodeling there was ongoing. Radiation physicist Mary Lou Curtis recalled the challenging working conditions: "We nearly froze the first winter, with only one small electric heater for our large lab. In the summer, the heat was stifling. Our windows were sealed shut, I presume for security reasons," she said.[75] A cafeteria at Unit I on Nicholas Road fed Dayton Project workers, some of whom shared a ride there in the company station wagon. "We were piled several layers deep, and a spirit of camaraderie soon developed," Curtis recalled. On Saturdays, however, the Unit I cafeteria was shut, so the female staff at Bonebrake took turns bringing in lunch, which they ate in the ladies lounge. So little was known about radiation and health safety, that the women didn't think twice about eating lunch near the laboratories. "All was well until one Saturday when the girls in charge brought in waffle irons and made waffles for lunch," Curtis said. "The smell of baking waffles wafted through the entire building. At this point, Dr. Lum explained to us that since little was known about the effects of ingested radioactive materials, it would be better if we did not eat in the building. He assigned the station wagon to us to go wherever we liked for Saturday lunches."[76]

Thomas's update on the overall plutonium work to Groves and Conant in November stated that chemists in Los Alamos and Chicago were perfecting the methods of analysis of "49," the code name given to plutonium (element 94–239). He also detailed the need for work on specifications to ensure delivery schedules and the final form of the product, sources of supply, processes for extraction, and concentration and stabilization of the final product. He reported that during his October visit to Los Alamos, Oppenheimer had set a tentative polonium delivery schedule.[77] Even with lead dioxide processing under way and delivery schedules set, Dayton scientists continued to iron out problems with the method. The quality of the first shipment of lead dioxide from Port Hope was not what had been promised, and less polonium than desired could be obtained from the residues.[78]

Oppenheimer met with Thomas on December 18 to review the lead dioxide process. He then shared his misgivings about it in a letter to Groves, pointing to irradiated bismuth as a better source for polonium: "You will note that the situation with regard to the residues is rather unsatisfactory and that at the present time it is by no means sure that the residues can be adequate for the future needs of the project. For this reason we have recommended that provision be made, as was I believe originally contemplated, for the irradiation of bismuth in the W pile."[79] To troubleshoot the quality of the lead dioxide residues, Lum, Fernelius, and Col. John R. Ruhoff, director of the MED's Materials Section, were sent to the radium refinery in Canada from December 21 to 23.[80] With a short break for Christmas, the trio then traveled on December 31 to Firestone Metal Products Co. to discuss a process used by that company for obtaining polonium for spark plugs.

By January 1944, the "first stage" of the Manhattan Project's chemistry program—"perfecting methods of analysis to be performed when sufficient quantities of 49 are available"—was complete, and the focus shifted to final purification.[81] Staffing issues, which involved the shuffling of chemists and metallurgists from Chicago and Berkeley to Los Alamos, were also nearly settled. Kennedy was officially appointed director of Los Alamos's chemical and metallurgical division, with Cyril Smith named associate director. Thomas noted that the personnel transfers that would soon have Los Alamos adequately staffed were creating problems with morale. "The recommended transfers have caused some uneasiness on the part of the remaining groups at Chicago and Ames as to just what their work is to be in the final purification and metallurgy," he wrote to Groves and Conant in his January progress report.[82] Chicago and Ames, he clarified, were to focus on theoretical work, while Site Y should concentrate on the applications to the final step, he wrote. The chemical and metallurgical program would be discussed at the monthly Chicago meeting on January 13 and "revamped" for the final stage.[83] A Monsanto internal report listed Dayton's activities in simple terms: "a secret war project which cannot be disclosed at this time."[84]

CHAPTER 8

POLONIUM PURIFICATION

AS WITH MUCH of the science undertaken during the Manhattan Project, the chemists in Dayton explored many processes simultaneously as they raced to solve scientific challenges and meet the bomb deadline. Some routes of inquiry led to dead-ends and some to success as the chemists worked to unlock the secrets of polonium purification. Ultimately, the scientists in Dayton pursued three methods for recovering polonium from lead dioxide: the first two methods were a wet process and the third was a dry, or kiln, process. In all cases, the recovery processes involved many steps. In some cases, they required huge tanks for immersing, rinsing, and storing the polonium.

THE DILLON METHOD

To ensure adequate quantities of polonium, the Dayton scientists experimented with a process used by Firestone Tire & Rubber Company in the manufacture of spark plug electrodes that was known as the Dillon Method, after its inventor J. H. Dillon. This was a lead chloride process in which polonium was concentrated on metal foils by immersing them in saturated solutions of lead chloride, which had been prepared by treating active lead dioxide with hydrochloric acid.[1] The polonium was then deposited on nickel plates in the acid solution, and the lead chloride precipitate was stored in crocks for six to 12 months, at which time the plating operation was repeated to retrieve additional polonium formed through the lead decay.[2] Firestone collaborated with Monsanto to convert nine tons of lead dioxide to lead chloride and plate the polonium on nickel foils. Monsanto's Plant B in Sauget, Illinois, was chosen for this production, as it had the necessary large, glass-lined equipment. Thomas had predicted delivery from this method of 9 curies of polonium on nickel foil by early February from Monsanto and later in the month from Firestone. By March 1, 1944, however, only three tons of radioactive lead had been processed, with about 2.5 curies of polonium deposited on nickel and copper plates.[3] The polonium was spread over 461 metal plates, covering a total surface area of about 1,700 square feet. Attempts to concentrate and collect polonium from the metal surfaces by distillation in a large still were only partially successful. When other processes were shown to be more effective, the Dillon process was abandoned.[4]

THE DRY METHOD—VOLATILIZATION

The dry (kiln) method focused on the volatility, or tendency to vaporize when heated, of polonium. It was surmised that polonium might be separated from lead dioxide by direct heating; the alpha and beta activity could be removed from lead dioxide by heating the material to 600° to 700°C in a stream of carbon dioxide.[5] In January 1944, a small test kiln was set up at Unit III to process 8-pound batches of lead dioxide by distilling polonium from metal foils and collecting the volatized element on cooled aluminum strips. The scientists built and tested stainless steel kilns—an agitator type with a capacity of 3 to 8 pounds and a rotary type with a capacity of 15 pounds—through which vaporized polonium flowed countercurrent to the cold lead dioxide. Each was designed with dimensions proportionate to a proposed full-scale production model. The agitator kiln failed, because the charge formed a mass at 700°C and resulted in slag that had to be chiseled out of the kiln. Lead orthophosphate (made by treating lead dioxide with phosphoric acid) was then used in place of lead dioxide. It required heating for four hours at 700°C, with collection in a Pyrex glass-wool filter condenser. Results from the kiln process were inconsistent, because it was difficult to filter out dust particles and to prevent fine dust from carrying the active material through the filters. The rotary kiln operated well mechanically, but it produced discouraging volatilization and collection data. Excessive dust covered nearby equipment and personnel with contaminated material. The kiln method was abandoned.

THE WET METHOD

The wet method, which involved chemical treatment of lead dioxide, proved to be the most successful of the approaches but required more space. The lead dioxide was first converted to lead nitrate by treatment with concentrated nitric acid and 27 percent hydrogen peroxide. Next, lead carbonate was added to reduce acidity to a pH level of approximately 4.0. The lead carbonate slurry caused a number of undesirable elements, including iron and aluminum, to precipitate out. Eighty pounds of slurry was required for each batch of 1,200 pounds of lead dioxide. The lead nitrate mixture was placed in a settling tank, where it was then filtered into a series of tanks. The residue separated by filtration was a complete mixture that carried with it the polonium that had grown in the lead dioxide prior to treatment.

Richard Yalman, a Harvard-educated chemist recruited by his former Ohio State doctoral advisor Fernelius, joined the project with the SED in May 1944 to work on the wet chemical process. Yalman recalled the work: "I didn't know what they were called at the time. All I knew was we had these small objects that we were to electroplate polonium onto. We used to joke about them. One time we had an order for fifty by a certain time. We'd joke and say, 'Well, they're going to General Groves so he can light up his desk at night,'" he recalled, referring to polonium's blue glow.[6]

Yalman partnered with chemist Henry Kuivila at Unit III on processes for isolating polonium from the lead dioxide residues. "We developed a process which was accepted, and when Unit IV was made, the process for isolating polonium involved getting it in a solution from ore residues, and putting it into large tanks, what we called milking the tanks for the autoclave," Yalman recalled.[7] The milking process, done every 60 days, was based on the finding that lead remains in the lead dioxide by-product after the uranium and radium have been removed from the ore. The lead transmutes first into bismuth and then to polonium.[8] Unit IV contained ten 10,000- or 20,000-gallon tanks where the radium-D solution and the polonium were held, Yalman recalled.[9] A total of 16 pilot plant runs with approximately 70,000 pounds of lead dioxide took place at Unit IV by this method, producing several curies of polonium.[10]

THE ARRIVAL OF X-10 PLUTONIUM AND BISMUTH

Clinton's X-10 reactor had gone critical in November 1943—about the time the Dayton Project began to organize—and at the end of the month produced plutonium. "The beginning of 1944 finds our Project deep in the problems of plutonium production, extraction, and purification," Seaborg wrote in his journal. "This vast involvement with a secret, synthetic element unheard of not much longer than two years ago and unseen until sixteen months ago in August 1942, would seem incredible to the outside world. Moreover, the means of producing plutonium in copious amounts—the chain-reacting pile—became operational just one year ago."[11]

In January 1944, as Clinton began providing plutonium for experimental use to scientists at the Met Lab and at Site Y, it also began sending samples of irradiated bismuth for polonium separation and purification to Dayton. The Clinton material contained from 0.032 to 0.083 curies of polonium per kilogram of bismuth. On a larger scale, 10 pounds of bismuth irradiated for 140 days was estimated to give 1 curie of polonium.[12] While polonium purification could be handled at Unit III on a laboratory-scale, the scientists needed a larger facility for full-scale production of lead dioxide processing and for the new work with the 110-pound batches of bismuth.[13] The only space that was both available and big enough was, coincidentally, the Talbott family's Runnymede Playhouse. The Dayton site also needed additional

8.1. Dayton's Oakwood area. Runnymede Playhouse (Unit IV) can be seen at the upper left, to the right of the grassy clearing. Mound Science and Energy Museum.

personnel for the increased effort, so staff was expanded from 17 to 33 chemists and chemical engineers.[14]

Katharine Talbott, Thomas's mother-in-law, had died in 1935, and her mansion had been torn down two years later, but the family's recreational Playhouse remained on the Runnymede Estate and was still in use by the extended Talbott clan. Thomas's teenage son used its greenhouses as an indoor victory garden where he grew tomatoes. Negotiations for the property were complicated by the interests of various stakeholders—the Talbott family owned the property; Oakwood City Council wanted it for a community center and protested that zoning laws prohibited the proposed commercial activities; and the Army wanted it for the bomb project. Condemnation proceedings were instituted under the Emergency Powers Act in a declaration signed by Secretary of War Henry L. Stimson. The U.S. Army Corps of Engineers seized the 3.8-acre property at 715 Runnymede Road by

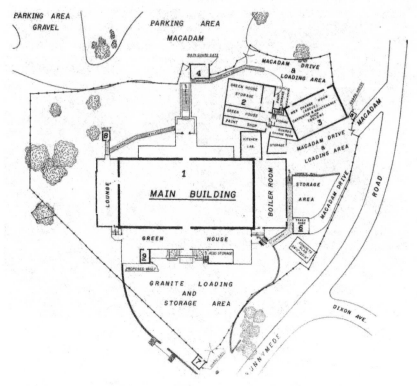

8.2. Unit IV site map, c. 1947. U.S. Department of Energy/Office of Classification.

eminent domain, and a lease was signed between the War Department and Talbott Realty Company on March 10, 1944.

The Playhouse was rented for $4,266.72 annually, with the stipulation that the building was to be returned to the Talbott family after the war, in its original condition.[15] The contract was initially drawn up through June 30, 1944, and then extended annually through June 1949. It stated that the building would be used as a training film production laboratory for the Army Signal Corps. It was, in reality, to become Unit IV of the Dayton Project and a polonium processing facility.[17]

CHAPTER 9

POLONIUM IN THE PLAYHOUSE

WORK ON CONVERTING Runnymede Playhouse at Runnymede Road and Dixon Avenue into a top-secret laboratory and processing facility began on February 15, 1944. A fence and three guardhouses were added to the property.

Indoors, the tennis court was subdivided into several smaller rooms and the ceiling dropped. Heating, air-conditioning, and air filtration ducts were run between the dropped and original ceilings. Balconies around the stage became radiation-counting laboratories. One of the greenhouses became a loading dock. As with other parts of the Manhattan Project, facility construction costs were a sizeable portion of the money spent on the Dayton Project. Total Manhattan Project expenditure for the Dayton Project was $3,866,507.00, of which an estimated $450,000 was dedicated to remodeling, installation of process equipment, and construction of buildings on the Dayton sites.[1]

John E. Bradley, a member of the SED from Oak Ridge who was assigned to Dayton as a health physics worker, recalled his impression of the Playhouse before it was converted to a processing facility: "It had a rubber tree, squash courts, statues and an indoor tennis

9.1. The Playhouse, secured by a fence and guardhouses for the secret work. National Archives (Atlanta), Records of the Atomic Energy Commission.

court. That was quite some place for a lab."[2] Floodlights illuminated the barbed-wire fence that surrounded the property, much to the annoyance of neighbors. Newspaper reports indicated that residents wondered what was happening, but a 24-hour guard was maintained to keep the curious out—43 armed guards patrolled Units III and IV.[3] To appease the neighbors, "careful consideration was given in order to minimize annoyances such as noise, smoke and dirt so as not to incur undue criticism from the residential area."[4] The wartime credo of "don't ask, don't tell" worked. "Between big trucks rolling in and out, the floodlights and heavy-duty power lines strung all over, the place was a real mess. But those were the days when you knew enough not to ask questions," recalled Oakwood resident Lee Jones.[5] "I walked by the area often, and knew something was going on, but it was very hush, hush," said former resident Joe Dooley. "I knew some kids whose parents worked there, but they said nothing. There were a number of 'secret' things going on around Dayton. We found

9.2. The Playhouse interior, as it appeared before conversion into a polonium-processing facility. Photo courtesy of the Talbott family.

out about them after the war."[6] After the war, a newspaper report described the site as something like a thriller movie: "Neighbors had termed the Playhouse a nuisance. They claimed white-clad figures walked through the grounds at night, that choking vapors filtered from the laboratories and that the area was radioactive."[7]

In February 1944, with work under way, Conant commented on the progress of the bomb project's overall chemistry and metallurgy division in a letter to Thomas: "I have no suggestions to make in regard to this whole chemistry and metallurgical program, which seems to me to be shaping up in excellent fashion. I take it that a great deal depends on the results which are going to be available before long and for the first time will give us the details of the problems which we are really up against."[8] By April, the Thomas progress report to Groves and Conant indicated that the purification, metallurgy, and chemistry of plutonium were in a "transition" phase that was causing a few glitches in the work. The problem lay in the difference between laboratory-produced plutonium and that produced in the Clinton pile. Lab-produced plutonium had a specific gravity of 16, while the material produced at X-10 had a specific gravity of 20. "Until now, the work had been carried out with microgram quantities; now that gram amounts of 49 are available, the work is being repeated and the results do not always check those obtained on a micro-scale. The fact that the work with larger amounts of 49 shows that we must revise some of our ideas is responsible for a certain tension and restlessness which I believe will disappear as we become surer of our ground," Thomas wrote.[9]

Thomas, who was known for his optimism, assured the Manhattan Project leaders that the glitches could be overcome. "I am confident that if the various groups stick to their respective programs we shall iron out the discrepancies arising in going over from microgram amounts to larger quantities of 49," he wrote.[10] His confident outlook was also apparent in a communication to Groves about the polonium work in Dayton:

> To be sure of an adequate supply of polonium, we have been pushing both sources, looking on the lead residues as an alternate source. With lead dioxide the only known method was a long and laborious process of extracting the polonium, very much as the Curies did at the beginning of the century. Accordingly we began research to develop a process which would enable us to recover polonium with modern chemical engineering methods and at low operating expense. . . . I am happy to report that this has been accomplished by our so called calcining method in which polonium is volatilized from the lead residues and collected on a suitable condenser. So we are in the fortunate position of having both horses come through.[11]

POLONIUM OPERATIONS BEGIN

On June 1, with renovations at the Playhouse complete, the Unit III polonium production group moved into the converted recreation center. Due to the central importance of the initiator work, security at the Playhouse was tighter than at other Manhattan Project sites. Even within the Manhattan Project, few people acknowledged the work in Dayton. "It was considered very secret," recalled Kenneth Nichols, who served as the Army's Manhattan District Engineer overseeing uranium and plutonium production at Oak Ridge and Hanford. "It was part of the weapon (but) we gave very little indication to anybody we needed it."[12]

Personnel records reflect the growth of the Dayton Project, with a significant increase in the number of new hires in 1944. Among Dayton's new chemists were Lowell V. Coulter (PhD 1940, University of California, Berkeley), who arrived from Boston University; Louis E. Marchi (PhD 1942, Ohio State), who was on leave from the Indiana University chemistry faculty; Fred J. Leitz (PhD 1943, University of California, Berkeley), who had researched radiochemistry and bismuth extraction processes at Stanford University; Robert A. Staniforth (PhD 1943, Ohio State), from General Aniline & Film Corp. with a knowledge of nuclear chemistry; Henry G. Kuivila (MA 1944, Ohio State); and John W. Schulte.[13] Some scientists had a "pretty good idea of what was being built and knew it was going to be big," said chemical engineer Edward McCarthy, who arrived from the U.S. Rubber Company in January 1944 and later served as director of the Atomic Energy Commission's Mound Laboratory. Most of those joining the project, however, knew little to nothing about the overall product.[14]

Staff in Dayton doubled from 46 full-time employees at the end of 1943 to 101 by the end of 1944. As the polonium work intensified for final delivery, that number grew again to 201 at the end of 1945 and had reached 334 by the end of 1946. An additional 34 or so men assigned by the Army's Special Engineer Detachment served Dayton during 1945 and 1946.[15] Of the 334 on payroll in 1946, 49 people were in research and 68 dedicated to operations; the remainder of the staff were in administration, service, and maintenance positions.

Samuel S. Jones was a chemistry graduate student at Cornell University when he was told to enlist in the Army and either become a combat soldier or join a secret project. Rather than risk being sent to the front lines, Jones chose the latter and ended up with the SED in Dayton, first at Bonebrake and then at the Playhouse. He had no idea there were other research facilities around the country working to create the atomic bomb. "Each Manhattan Project facility was kept secret from the others," he said.[16] The tight secrecy of the project led to morale problems; Dayton scientists were invisible to others in the

Manhattan Project. "When one of our people would go somewhere to meet somebody, they were told that our work wasn't significant. That produced morale problems we had to fight all the way through," Fernelius said.[17] The issue lasted long after the war. J. C. Franklin, manager of Oak Ridge Operations, referred to the continuing cloak of secrecy in a 1949 letter to Hochwalt: "Unfortunately, security considerations prevent our giving you this recognition and commendation publicly."[18] In 1983, George Mahfouz, who began work as a process engineer at the Playhouse in 1946, expressed the sentiments of many Dayton Project veterans when he recalled his disappointment at not receiving any credit: "For a while my nose was out of joint. When you figure the amount of work that went into this thing, you'd like to get some recognition."[19]

KEEPING AN EYE ON PERSONNEL ISSUES

Thomas, who was so occupied with his duties coordinating Manhattan Project chemistry and metallurgy that he rarely set foot in the Dayton laboratories, was keenly aware of personnel issues not only in Ohio but also across the Manhattan Project. He mentioned this to Groves and Conant in a June 1944 letter, reporting that "personnel issues are always with us but they have greatly decreased during the last 60 days."[20] Some of the problems were related to scientific territoriality and a clash of cultures. Groves viewed the scientists as "long hairs" and prima donnas who had their heads in the "clouds." The physicists, mathematicians, and chemists viewed one another warily. And most all resented to a certain extent the industrial engineers who translated their theories into large-scale production. Each occupied a corner of the Manhattan Project ring. "Thus in the Manhattan Project there were thrown together large groups of people from industry and from universities. They didn't understand each other too well," Thomas observed.[21] To illustrate his point, he shared an anecdote about meeting a well-known physicist during one of his trips to Los Alamos. The physicist inquired as to where Thomas taught physics. He responded that he was a chemist. "Oh, chemistry, one of the lesser sciences," the man replied. "This was quite crushing to me because on the previous day I had been talking to a mathematician who had given me to understand that all science was just a byproduct of mathematics," Thomas said.[22] He later noted that the role of the chemist and the chemical engineer had been underestimated in the Manhattan Project and that chemists played a critical role in developing the bomb. "Without the splendid work of the chemist supplementing and complementing the physical work, the bomb would not have been a success," he said. "As to numbers, there were several times more chemists in the Manhattan Project than there were physicists. This is jokingly referred to by the physicists as 'one physicist is worth two or three chemists;' but I do not look at it in this light. More correctly, I think it illustrates the late realization of the importance of chemistry to the project."[23] Rather than debate the relative importance of physics and chemistry, Thomas was quick to observe that the initial discovery of fission was made by a partnership between a radiochemist (Otto Frisch) and a physicist (Lise Meitner) and that cooperation between academia, military, industry, and government was essential to the success of the Manhattan Project.[24]

> I should like to take you behind the scenes of our atomic research and show that this cooperation was something that had to be worked for, all recognizing its importance. Naturally, it was not always a serene and happy family as some writers would lead us to believe. Cooperation was attained, people did learn to work together, with the result that the project itself did not fission, largely because of a few leaders like (Arthur Compton), Dr. Bush, Dr. Conant, General Groves and his staff. That in itself was an achievement.[25]

The secrecy of the Dayton Project meant that it was hidden from local authorities, who were unaware of the work being done within their jurisdiction. This caused confusion when they were presented with a Dayton Project worker's identification badge. Among Dayton Project recruits were some 30 or so SEDs. Although military men, they dressed in civilian clothing to avoid scrutiny. One day while off duty, Will Konneker, a member of the detachment who later became a pioneer in the nuclear medicine field, was stopped by a police officer for a minor traffic infraction. He was asked for identification, but his Class A pass showing special detached duty was not deemed adequate. Tight-lipped about his work, Konneker would not respond to the questions he was asked, so he was taken to jail. An officer at Wright Field, who had been designated as the contact man for identification of all SED men, couldn't be reached that night, so Konneker spent the night in jail.[26]

Even between individual Dayton labs, scientists did not know what their neighbors were doing. Betty Halley, who worked in the counting room, said there was little to no exchange of information between researchers. Despite all best efforts, security leaks did occur. They ranged from a scientist's wife being surprised when she learned that the other women at a card game didn't know what their husbands were doing to scientists themselves being loose-lipped, and to near-slips at higher levels. Arthur Compton, director of Chicago's Met Lab, visited Dayton early in the project and spoke with technical employees at Unit III. Mary Lou Curtis, who worked in the counting laboratory, was among those in attendance. "As his speech progressed he divulged that the work was in the nature of development of a secret weapon, 'we don't know how far Germany has progressed; but whoever gets the answer first will win the war.' As he reached this point, however, Dr. Lum, fearing a breach of security, rapidly changed the subject. This proved to be the biggest hint about the nature of their work that the Monsanto employees received until the bomb was dropped on Japan," Curtis said.[27] Following Compton's visit, word spread among the employees that they were working on a form of energy that would be used in tanks, Lum later recalled.[28]

POLONIUM DELIVERY GETS UNDER WAY

On February 24, 1944, Thomas notified Oppenheimer that Dayton had approximately 2.0 curies of polonium in solution and expected to plate it on a foil by March 1.[29] He noted that it was somewhat impure and could either be improved before shipping or sent to Los Alamos in its current condition.[30] Oppenheimer requested 50 percent polonium and delivery by mid-March.[31] Accordingly, the first shipment of polonium produced from the ore residues left Dayton for Site Y in a lead-lined suitcase on March 15. The shipment contained 0.21 and 0.35 curies of polonium on foils and two vials of polonium in solution, one containing 1.35 curies and the other 0.5 curies. Though seemingly miniscule amounts, this quantity emits thousands of neutrons per second, enough to trigger the device. The solutions were in sealed glass vials, which were contained in glass-stoppered vials filled with glass wool. Transport notes from Oppenheimer recommended, "No special atmosphere necessary for first shipment except that it be reasonably dust free. Suggest foil or foils be shipped in all glass container, e.g. wide mouth weighing bottle, packed in such a way that deposit cannot rub against anything. This container should probably be packed in glass wool to simplify recovery in event of any severe accident during transportation."[32] One of the vials—containing 1.35 curies of polonium—shattered en route, but the glass wool captured the polonium. It was returned to Dayton, recaptured, and reprocessed.[33]

By April 1944, Thomas had good news for Groves and Conant: "The development of . . . processes for extraction and concentration of polonium is largely completed."[34] Lead dioxide processes were under way and irradiated bismuth looked promising. One ton of the first irradiated bismuth contained from 29 to 75 curies of polonium, a notable increase over the yield of 1 to 3 curies per ton from the Port Hope lead dioxide.[35] By August, Site Y had 6 curies of polonium on hand, enough that chemistry group leader Richard Dodson was able to request that 2 curies be set aside from each monthly shipment for research.[36] It seemed that polonium production and purification was in good shape. So, too, was plutonium production.

PILE-PRODUCED PLUTONIUM POSES PROBLEMS

In order to yield 1 gram of plutonium per day, the Clinton reactor was required to generate 500–1,500 kilowatts (kW) of energy.[37] X-10 had surpassed this level by May 1944, when it was operating at 1,800-kW; by June and July it had reached 4,000-kW. On June 13, the plutonium production situation looked promising enough that Thomas sent Groves and Conant a note that exuded confidence: "As the news is good, this report will be brief. . . . The first phase of our problem, namely, obtaining metal of 98% purity for the implosion method, has been reached in our laboratory."[38] Groves was so encouraged that he reported to the Military Policy Committee that progress was "quite satisfactory and fully up to expectations." Richard Hewlett, chief historian of the Atomic Energy Commission, later commented: "It seemed that one of the major objectives, pure plutonium, had been all but attained."[39]

Although production quotas had been met, the science of plutonium was still uncertain. Until January 1944, when X-10 had begun producing enough plutonium for experimental use, chemists did not have enough material to study plutonium's chemical and metallurgical properties. In April, with samples from X-10 for experimentation, Segrè's radioactivity group in Los Alamos confirmed earlier suspicions that Clinton-produced plutonium contained impurities that threatened pre-detonation. Segrè discovered that the impurities came from plutonium-240, a light-element impurity by-product created in the reactor production of Pu-239. Plutonium-240 is a prolific emitter of alpha particles.[40] When the alpha particles strike nuclei of light element impurities such as beryllium and aluminum, neutrons are generated. The problem with Pu-240 was that the spontaneous emissions of neutrons would cause a premature detonation, or fizzle, that threatened the slow-detonating gun assembly design under consideration for both the uranium and plutonium bombs.[41] The scientists anticipated that the plutonium from Hanford would have even higher levels of impurity, due to the higher neutron flux in the reactors there.[42]

On July 17, Oppenheimer traveled to Chicago to share the discouraging news with the Chemistry and Metallurgy Project Advisory Board during its meeting in Room 209 of the University of Chicago's Eckhart Hall.[43] The information he reported seemed a near-catastrophic blow to the Project chemistry undertaken to that point and would require what Emilio Segrè termed a "reorientation of the program."[44]

The change in direction would greatly affect the Met Lab plutonium purification team, which was led by Seaborg. "The program was scheduled to discuss (a) postwar plans for the Project as a guide for present changes in Project policy and organization; (b) the importance of light water moderated units in the overall Project program and the effect on transfers of associated personnel required if the program is to be pushed. The meeting, however, concentrated on something that came up which was much more immediate," Seaborg wrote in his journal. "Robert Oppenheimer, who was attending the Board Meeting from Los Alamos, announced that E. Segrè, S. O. Chamberlain, and G. W. Farwell have found strong evidence for the existence of the plutonium isotope ^{240}Pu, which undergoes decay by spontaneous fission! This was found in the neutron-irradiated ^{239}Pu from the Clinton pile that we had purified for them."[45]

Seaborg went on to describe the impact of the news: "It should be noted that this disclosure came as a great shock to everyone. . . . Because of this new development, Site Y will now have to rethink how it will proceed in the design of a plutonium bomb."[46]

Oppenheimer relayed the news to Groves the next day: "The fission rate of the sample appears to be proportional to the number of neutrons which have previously passed over the material," he wrote. "Extrapolation of these results to Hanford product would indicate a neutron emission of several hundred times that permitted in setting up the chemical specifications."[47] He recommended discontinuing the plutonium-purification gun assembly program and turning, instead, to a plutonium implosion bomb, which would rely on plutonium produced at Hanford.

Seaborg was notified of the advisory board's decision on July 19. "(T. R.) Hogness and (John C.) Warner came to my office to tell me that the purification program is no longer needed. I was standing in the hall in front of my office as they approached. They said that Compton had agreed I should be given the reason but that I was not to tell others. I said they didn't need to tell me the reason—I assumed the spontaneous fission rate of ^{240}Pu has been found to be so high as to overshadow the neutrons from the α,η reaction.[48] I went on to say that, since no one has given me this information, I feel free to pass my interpretation on to my men," he reported in his journal.[49] "Compton took up in more detail the decision reached at the meeting of the Project Advisory Board Monday night, namely, that the need for exceedingly high purity plutonium no longer exists and that the present intensive work on purification can be dropped. Hogness asked if it would be proper to inform others on our staffs that it is the properties of plutonium that have made the change in emphasis necessary; adding that he was 'thinking especially of Seaborg.' Compton said he believed I should know. The section chiefs at Argonne (a second pile built west of Chicago) also should know, he added, but it is highly essential to limit the information to the smallest number of people. Fermi mentioned the implications of such a finding must be known in order to plan properly for the construction of new piles."

The news was a blow to the project and put into question the massive and expensive plutonium production site at Hanford. Was there a need to continue the intensive effort to purify plutonium if the rogue isotope would nonetheless be present?[50] Groves arrived in Chicago on July 20 to discuss the matter in person. The next day, he conferred with Oppenheimer, Thomas, Conant, and Compton, and made the final decision to abandon the plutonium gun-assembly method, believing that the ongoing quest for plutonium purity would delay the project for too long.[51] They opted instead to pursue a plutonium implosion bomb, nicknamed the "Gadget," that would use a polonium-beryllium initiator known as the "urchin" and was less sensitive to impurities. Conant communicated his distress by writing across Thomas's June 13 report: "All to no avail, alas!"[52]

The change in direction marked the beginning of the second half of Manhattan Project history. For the chemistry and metallurgy division, it also meant a change in staffing. From this point on, many of the chemistry personnel at the Met Lab were reassigned to other sites across the Project, including Clinton, Hanford, and Los Alamos. Research efforts had achieved a purification level that was acceptable for the new work.

Thomas informed Groves that he was disbanding the plutonium purification staff. "It seems unnecessary for the military objective to continue the program on final purification. I am proceeding with the demobilization of my staff which was handling the coordination and general direction of the chemistry, purification and final metallurgy of 49," he wrote.[53] As Seaborg explained in his journal, "the planned extreme purification of plutonium would be futile—this could not prevent the emission of the unwanted neutrons."[54] Thomas told Groves that the monthly Chicago meetings would be discontinued, and that his final act as coordinator of the plutonium program would be to prepare a detailed document on the chemistry and metallurgy of 49.[55]

Work on that document got under way on August 3, when Seaborg and his deputy director, John C. Warner, arrived in Dayton to work with Thomas, Hochwalt, and Robert Staniforth.[56] The final

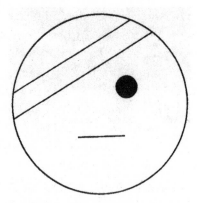

9.3. The alchemical symbol for plutonium designed by Charles Allen Thomas, who doodled it on a napkin during a meeting. Courtesy of the Thomas family.

document, *The Chemistry, Purification, and Metallurgy of Plutonium,* was issued in January 1945 as two volumes totaling 539 pages. Included in the text was the alchemical symbol for plutonium, which Thomas had reportedly first sketched on a napkin during a dull meeting.

POLONIUM AND THE IMPLOSION METHOD

Although the direction of plutonium chemistry had been temporarily thrown into question by the confirmation of Pu-240's spontaneous fission risk, results from Latimer's lab in Berkeley provided sufficient proof of the potency of polonium to enable the scientists to shift the focus to the implosion device, which would use a polonium initiator to trigger the plutonium bomb.[57] "It was decided to attack the problem of the implosion with every means available, to throw the book at it," reports the Manhattan Engineer District history.[58] The Project was reorganized in August 1944 for the effort.

The initiators, known as "urchins," were spheres about 0.8 inches diameter that contained interior cavities lined with teeth that projected into a hollow center. Polonium and beryllium were deposited on opposite sides of the teeth. Upon being crushed by the incoming projectile piece of fissile material (uranium bomb) or by an implosion (plutonium bomb), the polonium and beryllium would mix: alphas from the polonium would strike beryllium nuclei and liberate neutrons to trigger the detonation. Each initiator used about 50 curies of polonium-210. The polonium was to come from bismuth irradiated in the X-10 experimental reactor at Clinton and Hanford.[59] It would then be shipped to Dayton, where it would be purified. Whether Hanford and Dayton could produce and purify enough polonium for a plutonium implosion bomb was the unknown; it was Monsanto's job in Dayton to answer that question.[60] In the meantime, Oppenheimer, Groves, and Conant had other duties for Thomas in mind—they needed him in Los Alamos.

At a breakfast meeting with Thomas during their visit to Chicago on July 17, Oppenheimer and Conant asked Thomas to move to Los Alamos for a two-month period to help smooth out operations there as the project underwent significant reorganization. Thomas evidently asked for time to consider the request, because Conant wrote him 10 days later, asking for his help: "I don't believe you can realize how much it would be possible to accomplish by such a move, or how desperate the situation is from the overall point of view in regard to what happens on this front in the next four months." Conant added that his note was "just one more urgent appeal for you to examine your business and personal affairs to see if it would not be possible to take two months off and go out and lend a hand on this matter. Frankly, I can think of no move right now that would do more towards helping to shorten the war."[61] Conant, conspiring with Oppenheimer, was not at all sure he could convince Thomas to move, but he kept pushing. "I am writing to Thomas today to reinforce your appeals. I do not know what success I will have," Conant wrote to Oppenheimer on July 27.[62]

Thomas once again declined to move to Los Alamos. "The extreme seriousness and importance of the job at Y is fully recognized, but I am not at all convinced that I could be a factor in effecting a final solution to the problem. I say this because it is entirely out of my field.... Against this has to be weighed the importance of improving and finishing some developments for Monsanto that are actually being utilized on the fighting front, where there is no doubt of my usefulness. Also, there is the job of completing the final stages of polonium production, overseeing the enlargement of the propellant facilities here in Dayton, and compiling the final report on the Chemistry and Metallurgy of 49," Thomas wrote to Conant. He promised, though, to spend a week in Los Alamos meeting about implosion with Oppenheimer and chemist and NDRC explosives expert George Kistiakowsky, who he had helped recruit to the Manhattan Project in February 1944 to oversee implosion work.[63]

With the bismuth-phosphate process conclusively chosen for mass polonium production, workers in Dayton quickly converted the Playhouse laboratories for the new work. Small laboratories replaced the larger rooms that had been used for lead dioxide processing. Equipment that had been used for the lead dioxide process was dismantled and shipped to Clinton.[64] The remaining lead dioxide was precipitated as carbonate slurry and shipped to the Madison Square Area Engineers Office in New York, which was to dry it and return it to the Canadian Government, from whom it had been leased.[65] Between November 1943 and May 1945, Dayton had processed a total of 37 tons (about 70,000 pounds) of lead dioxide, resulting in about 40 curies of Polonium-210; a lot of raw material for a miniscule but powerful quantity of product.[66]

By 1945, the chemistry and metallurgy division in Los Alamos had evolved into even more specialized groups; CM-5, led by Clifford Garner, oversaw plutonium purification; CM-8, led by Eric Jette, dealt with plutonium metallurgy; and CM-15, led by Iral Johns, focused on polonium. Work in the Dayton lab focused on deciding what information about polonium should be known, studying literature to determine what was already known, and devising experimental procedures to obtain new information. In order to make and study the properties of polonium compounds, it was necessary to become trained in the new technique of ultra-micro or "gamma scale" chemistry dealing with miniscule quantities of polonium.[67] The scientists spent a large portion of 1945 learning the techniques.[68]

By late April 1945, Dayton prepared for the first irradiated bismuth shipments from Hanford. Thomas met in Chicago with Kennedy, Col. Matthias (MED director of Hanford) and a group from DuPont (Hanford operators) to agree on upper limits for impurities in the shipments.[69] Bismuth from Clinton's experimental pile was shipped to Dayton in brick-form, each brick measuring 12 inches by

9.4. Slug canning operation at the Atomic Energy Commission's Mound Laboratories. Mound Science and Energy Museum, Miamisburg, Ohio.

¾ inch by ¾ inch and weighing 58 pounds. The bricks were packed in individual wooden boxes that snugly fit each brick. Once at the Playhouse, the bricks were stored in a tile-lined cavity in the floor near the processing area.[70]

Hanford received raw bismuth from the American Smelting and Refining Company and canned it, or placed it in aluminum alloy cans or "slugs" measuring 4 to 8 inches long by 1.5 inches in diameter.[71]

The bismuth slugs were then bombarded for about 100 days, and shipped to Dayton's Unit IV by military courier. The couriers traveled often by train or motored by truck or van along the nation's bumpy roads at a mandated 35 mph.[72] Once at Unit IV, the irradiated slugs were stored in a steel safe, or storage cabinet with doors on the front- and backsides. Tubes of somewhat shorter length than the depth of the safe were positioned horizontally in the storage cabinet, and the space between the tubes and the inner cabinet wall was filled with

9.5. Chemist Harry Weimer, standing second from left, joins Dayton Project coworkers in a lunch break at the Playhouse (Unit IV). Photo courtesy of Robert A. Weimer.

lead. The tubes were large enough to receive slugs loosely, and the doors of the cabinet afforded access to the tube ends. Slugs were easily inserted or removed from either side of the cabinet. Each tube was provided with a removable lead plug that reduced radiation effects from the stored slugs. The slugs were also stored underwater to contain the radiation. An inclined storage rack was installed in the pool, provided with holes into which slugs could be inserted or removed by tongs. A periscope was installed in the pool so that any slug could be inspected and its identification markings could be examined without lifting the slug above the surface of the water.[73]

Dayton scientists pursued a variety of processes for separating polonium from bismuth and purifying the product. Polonium was initially separated from the bismuth slugs by treating the slugs with hydrochloric acid, which dissolved the aluminum can and resulted in an aluminum chloride solution. That solution was discarded, and the de-canned slugs were then dissolved in a mixture of nitric acid and hydrochloric acid or aqua regia. The polonium then required further concentration before being electroplated on platinum foil for handling. A second method, attempted in 1944, introduced a bismuth scrub and added steps that were repeated to concentrate the polonium. Many refinements on this process eventually resulted in the processing of more than 50 tons of radioactive bismuth at the Playhouse.[74] One of the greatest challenges of the Dayton Project was controlling the purity of the polonium, which was processed as a solution and electroplated for handling. Impure plating solutions caused trouble, requiring repeated plating and stripping of the electrode, sometimes as many as ten times. Between 1944 and the end of the war, the scientists pursued many methods in their quest for the highest concentration and purity of polonium.

Beginning in June 1944, work in Dayton introduced a bismuth scrub to the purification process, adding bismuth powder to the denitrated solution to act as a polonium scavenger. The addition of a second bismuth scrub further reduced and concentrated the polonium. Further refinements were made with the introduction of volatilization—conversion into vapor—of polonium from bismuth powder in a large furnace. The polonium was captured on the walls of quartz tubes and then retrieved by refluxing with acid. This method resulted in much greater purity and eliminated the need for acid leaching.

Dayton scientists also experimented with an alternative silver scrub process in which polonium was scrubbed on silver powder or silver foil. The polonium spontaneously deposited on the silver, and the silver and polonium were then dissolved, with the silver separated out by precipitation as silver chloride. Full-scale investigation of this method took place in February 1945, using a tower in which silver was placed and through which the solution was passed. The method produced the most concentrated solutions of polonium to that time but still contained too many impurities. Compounding difficulties, the tower had to be dismantled after each run, which made the process impractical in production work.

Dayton workers also experimented with the production of polonium-beryllium neutron sources by plating polonium on platinum or beryllium disks and sandwiching them between beryllium disks. As indicated in a letter from T. S. Chapman of the MED's Chicago Area to Oppenheimer, the neutron source work was challenging: "Monsanto does not appear to have much experience in the preparation of such sources and is encountering difficulty in filling the requests in accordance with specifications. It is suggested that you transmit to this office for forwarding to the Monsanto Chemical Company instructions for the preparation of polonium-beryllium sources in accordance with the best techniques known to the men under your jurisdiction."[75] The efficiencies were low, and the work was discontinued.[76]

As recorded in flow charts, the bismuth separation process evolved over time as steps were added and sometimes removed:[77]

March 1944
- Solution of bismuth metal
- Electrolysis of solution
- Repeated stripping of foils and replating
- Purity by visual appearance
- Assay by solution difference

June 1944
- Solution of bismuth metal
- Initial bismuth scrub
- Secondary bismuth scrub
- Electrolysis of solution
- Repeated stripping of foils and replating
- Purity assay by visual appearance
- Calorimeter assay of plated foil
- Neutron assay by Szilard-Chalmers

November 1944
- Solution of bismuth metal
- Initial bismuth scrub
- Tellurium process
- Electrolysis of solution
- Volatilization of foils and replating (replaces acid stripping)
- Purity assay by straggling effect
- Calorimeter assay of plated foil
- Neutron assay by comparison

March 1945

- Solution of bismuth metal
- Initial bismuth scrub
- Secondary bismuth scrub
- Volatilization from powder
- Electrolysis of solution
- Calorimeter assay of plated foils
- Gamma Assay of plated foils
- Neutron assay of foils by tub method

September 1945

- Decanning of bismuth metal
- Solution of bismuth metal
- Initial bismuth scrub
- Secondary bismuth scrub
- Volatilization from scrub powder
- Removal of silver from solution
- Assay by gamma count
- Micro-purity assay by weighing
- Electrolysis of solution
- Calorimeter assay of plated foil
- Gamma assay of plated foil
- Neutron assay of plated foil

CHAPTER 10

HEALTH PHYSICS AND A SOVIET SPY

AT THE BEGINNING of the Dayton Project, nothing was known about the effects of polonium on the human body, nor was there an established means of detecting its presence in the human system. Health safety protocols for working with polonium were developed alongside the theoretical and physical science.

Polonium is chemically more toxic than hydrogen cyanide. Although it emits alpha particles that are easily stopped by paper or skin, it can damage human organs if inhaled or ingested. In recent times, polonium has most famously been tied to the 2006 poisoning murder of Russian KGB agent Alexander Litvinenko, who had defected to the British MI6 intelligence agency and was allegedly poisoned by drinking tea laced with polonium. Colonel Nichols, who visited the Dayton site periodically, acknowledged that the assignment there was made even tougher by the element of polonium itself. He remarked that it was a "very difficult assignment, not only because it had never been seen before, but also because of the radiation involved."[1]

Although polonium is eliminated rapidly and does not settle in dangerous concentrations in bone, it migrates easily and is toxic if ingested or inhaled and could contaminate Dayton Project workers and equipment. From the start of the Dayton Project, radiation levels were measured in the area by trucks equipped with detection equipment. Air, soil, and water were sampled up to 75 miles away from the Playhouse.[2] Employees, who were exposed to significant radiation on a daily basis, were checked regularly in some of the world's first health physics facilities. Louis B. Silverman compiled the first set of "General Health and Safety Rules" for the Dayton Project in December 1943 as the project started. Workers were to wear rubber gloves, use a respirator when handling dry materials, wear special clothing, and refrain from eating or smoking in the laboratory. More extensive investigation into health safety was not made until early spring 1944. In February 1944, Thomas noted to Groves and Conant that "protection of personnel from the hazards of the lead dioxide and bismuth processes was placed on a firmer basis with the acquiring of actual radiometric analyses of the two materials." The hazard involved from the alpha radiation from polonium itself is still an unknown factor, he noted, adding that operations were organized to minimize exposure.[3]

Safety equipment developed by the Manhattan Project's health physics division included pocket meters to measure exposure on individual workers. One, about the size and shape of a fountain pen, was electrostatically charged in the morning and read at the end of the day. A record was kept of the readings at the time the meter was handed out and when it was turned in. The degree to which the instrument discharged indicated the total amount of radiation to which the worker had been exposed. Film badges were also used. These small pieces of film were worn in the worker's identification badge, with the films periodically developed and examined for radiation blackening. Other radiation detection instruments used across the Manhattan Project included "Sneezy" for measuring the concentration of radioactive dust in the air, and "Pluto" for measuring alpha-emitting contamination (usually plutonium) of laboratory desks and equipment. Counters were used to check the contamination of laboratory coats before and after the coats were laundered. At the exit gates of certain laboratories, concealed counters sounded an alarm when someone passed whose clothing, skin, or hair was contaminated. Additionally, routine inspections of laboratory areas were carried out by health physics surveyors.[4]

GEORGE KOVAL: SOVIET SPY

Among the Dayton health safety officers responsible for site inspections was a member of the SED named George Koval, who transferred to Dayton from Oak Ridge on June 27, 1945.

Koval was amiable, baseball loving, and Iowa born, and was assigned as a surveyor at Unit III. The position gave him free access to all laboratories and areas, as he swept the sites, or surveyed them, for radiation. Nearly six decades later—in 2007—Koval was revealed to have been a Russian spy whose fortuitous transfer to the Dayton site at the peak of the top-secret polonium work enabled the Soviets to build and explode their own atomic bomb years ahead of schedule.

The son of Russian Jewish immigrants who had fled the persecution of the czar and settled in Sioux City, Iowa, Zhorzh (George) Koval was born on Christmas Day 1913 and graduated from Sioux City's Central High School in 1929 at age 15. He was a member of the honor society and the debate team, an actor in school productions, and a popular student. In this sense, he was much like Charles Allen Thomas, the Dayton Project leader who as a high school student was active with his high school's debate team, had given an award-winning speech on the topic of the Bolshevik menace, and was a popular singer. Unlike Thomas, Koval was an avowed communist even as a teenager. When Koval's family moved back to Russia in 1932, their son went with them. He enrolled at the Mendeleyev Institute of Chemical Technology in Moscow, married fellow student Lyudmila Ivanova, and earned his degree in chemistry in 1939. Then, under the ruse of being drafted into the Soviet army, he disappeared. Sometime between then and his return to the United States in October 1940 aboard a tanker that docked in San Francisco, Koval trained as an agent for Russia's intelligence agency, the *Glavnoje Razvedyvatel'noje Upravlenije* (GRU). Once in New York, a U.S. citizen by virtue of birth, Koval registered with Selective Service and assumed deputy command of the local GRU cell, which operated under the cover

10.1. George Koval, U.S. passport photo, 1948. U.S. Department of Justice/ Federal Bureau of Investigation.

of the Raven Electric Company, a supplier to firms such as General Electric.

Koval registered for the draft on January 2, 1941, but received "occupational" deferment beginning in February 1942 and was inducted into the Army on February 1943. After basic training at Fort Dix, New Jersey, Koval was sent to the Citadel for a month and then admitted to a new unit, the Army Specialized Training Program (ASTP), which provided skilled enlistees with education and technical training. Under the auspices of the program, Koval attended the City College of New York, where he studied electrical engineering. Unbeknownst to his fellow students, he had already studied the discipline at the University of Iowa and the Mendeleyev Institute. They viewed him as a mature student (he was 30) who somehow never studied and was popular with the ladies. He never spoke about politics or the Soviet Union.

In August 1944, Koval and about a dozen other scientists were assigned to the Manhattan Project through the Army's Special Engineer Detachment, which sent him to Clinton Engineer Works. He joined the Instruments Unit of the Health Physics Department and had wide access to all work areas and employees, which allowed him to report back on the plutonium work. His Soviet handlers charged him with watching Clinton's polonium supply. In February 1945—coincidentally around the time that a final decision was made to use the initiator design using polonium-beryllium—Koval took his first leave and met with his Soviet handler "Clyde" to transmit information back to Moscow.[5] His information reached Moscow via coded dispatches, couriers, and the Soviet Embassy. Among the intelligence he sent was that Clinton's materials were being sent to another Project site in New Mexico. He reported on shipments of enriched uranium and plutonium to Los Alamos, and about the scale of the effort at Oak Ridge and how it was connected to facilities elsewhere.[6] After 11 months in Tennessee, Koval got a firsthand look at polonium when he was transferred to Dayton, taking a room in a boarding house at 827 W. Grand Ave. The transfer occurred despite MED security regulations that generally restricted transfers of non-manual workers from one Manhattan Project site to another.

At six-feet tall and attractive, Koval had an engaging personality and reportedly got along with everyone. Jim Schoke, a member of the SED who was in the instrument group at Chicago's Met Lab, was a frequent visitor to Dayton, where he trained health safety workers, including Koval, on how to use, maintain, and service the equipment. The Chicago group also collaborated with Dayton physicists on equipment design for the polonium work, including a 1945 redesign of the quartz-fiber Kirk-Craig balance for weighing a polonium plate and construction of a balance to directly weigh product foils. Schoke trained Koval to measure alpha rays. Polonium was deposited on platinum disks and then plated. The instruments counted the electrical impulses developed by the alpha rays ionizing in the air.

According to monthly Health Section progress reports issued by Monsanto Unit III, "G. Koval" was among the 26 Health Section staff members on the Dayton Project in 1945. The reports indicate his area of responsibility as "Survey Unit #3—George Koval." As a health physics surveyor, Koval had authorized access to all laboratories and clean rooms, more so than nearly all of the other scientists and other staff. "Routine surveys consisting of thirty or more spot checks and six air samples in each lab are taken every day," Koval's September 1945 report stated.[7] Koval may also have had occasional access to the polonium purification work in Unit IV, though another health worker and fellow SED, John E. Bradley, was assigned to that area. Koval and Bradley partnered on a special survey of the Unit III outside area, co-authoring a report issued October 27, 1945.[8] Koval remained in Dayton until January 1946, and was discharged from the Army on February 12, 1946, as a Technician/Third Grade at Fort Atterbury, Indiana. He was awarded a Good Conduct Medal, a World War II Victory Medal, and an American Theater Service Ribbon.

After his discharge, Koval returned to CCNY and earned a Bachelor of Electrical Engineering there on February 1, 1948. Soon after, he was ordered back to the Soviet Union. He was issued a U.S. passport on March 15, 1948, for the purpose of four to six months of foreign travel to France, Belgium, Switzerland, Sweden, and Poland as European sales representative for the Atlas Trading Company, a bogus claim. On October 6, 1948, he sailed on the USS *America* from New York to Le Havre. From there, he made his way back to Moscow.

Koval died in Moscow in 2006 at the age of 92. In a posthumous ceremony in November 2007, then Russian President Vladimir Putin awarded him the nation's top honor for meritorious service, the gold star as a Hero of the Russian Federation.[9] The Russian presidential proclamation stated, "Mr. Koval, who operated under the pseudonym Delmar, provided information that helped speed up considerably the time it took for the Soviet Union to develop an atomic bomb of its own."

Of the few Dayton Project workers still alive when Koval's spying came to light, only machinist Howard DuFour remembered having had contact with him. DuFour recalled that FBI investigators came to Dayton in the 1950s and began asking about Koval.[10] While other Manhattan Project spies such as Julius and Ethel Rosenberg and Klaus Fuchs were caught after the war, Koval apparently was not scrutinized for many years. The *Manhattan District History*, commissioned by Groves in 1944 and publicly released in 2013, reports that "during the entire history of the project up to 31 December 1946, there was no known compromise of classified information or damage to buildings, equipment, or material which could be ascribed to the lack of physical protection against espionage or sabotage."[11] Although the FBI assembled a dossier of some 1,000 pages on Koval, the matter was kept confidential for 60 years, until the announcement of his posthumous award. Koval's niece told a reporter that her uncle never talked about his atomic bomb work, espionage or otherwise.

BUSINESS AS USUAL

As far as Dayton Project workers were concerned, George Koval was just one of the health physics employees monitoring radiation around the laboratories. His presence was, presumably, business as usual. The chemists worked in fume hoods and wore three layers of gloves—surgical, thin cloth, and thick rubber—that they removed, along with their laboratory coveralls and shoe covers before leaving the facility. After washing their hands in a diluted hydrochloric acid solution, alpha-particle radioactivity was measured using a specialized Geiger counter developed by physicist John Sopka.[12] The maximum tolerance allowed those leaving the building was 1,000 counts per minute per hand.[13] Dayton Project veteran George Mahfouz recalled a female co-worker who used bobby pins to tame her hair. She inadvertently contaminated herself, because she frequently held bobby pins in her mouth as she adjusted and reset her hair. She reportedly had the highest count of anyone at the Playhouse.[14]

Preliminary work in health physics showed that detection and estimation of radiation exposure could be made from urine by counting the disintegration of polonium on a copper disc. Early in 1944, a bioassay method was implemented for monitoring personnel and was reviewed in April, at which time tolerance levels were established. Routine urine counts were made twice weekly on all who handled polonium. Chemist Yalman recalled taking large flasks home on the weekends for 24- and 48-hour urine samples. His wife grew curious about the samples and went to the library to look up which chemicals might show up in urine; she learned that it might be radium.[15] Radium, she read, was contained in the ore sources of lead dioxide residues used for recovering highly radioactive polonium. She had inadvertently learned about her husband's work.

Dayton scientists and operation personnel wore coveralls or a laboratory smock, rubber gloves, a cap, respirator, shoe covers, and often a face shield. All of these items were worn only once and then laundered at Unit I.[16] Laundering created a substantial amount of work across the Manhattan Project. During the month of June 1945, the Los Alamos Chemistry and Metallurgy Division sent 23,098 pieces of protective clothing, 11,027 rubber gloves, and 564 respirators to the Site Y laundry facility.[17]

Personnel were also monitored using dental X-ray film badges, wrist badges, and finger ring badges to measure beta and gamma radiation; the badges were exchanged weekly. Each person also received two pocket ionization chambers sensitive to photon radiation. Beginning on February 22, 1944, when exposure limits per workday and workweek were established, some 206 whole body film badges were issued to Dayton workers by the University of Rochester office of the Manhattan Engineer District. When tolerance levels were exceeded, the worker in question was removed from the task or area where the exposure occurred. Workers who were identified as "hot" were asked to stay on the job site and sometimes passed the "cooling down" time by playing cards in the area that had previously served as the Playhouse lounge.[18]

At Unit IV, a thick, waterproof paper covered the floors. Health workers laid down fresh paper daily and secured it with masking tape. They swept it with instruments throughout the day, checking on radioactivity. If it became hot, they blocked off the area, peeled off the paper, appeared with a huge roll of fresh paper and installed a fresh floor cover.[19] The radiation was measured in the Counting Room at Unit III, where Mary Lou Curtis was one of several women who measured samples of radioactive materials under the supervision of physicist Sergio De Benedetti. The scientists were told not to mention him by name, because he had worked on radioactivity at the

10.2. Physicist Sergio De Benedetti at the Bartol Research Foundation of the Franklin Institute, Swarthmore College, c. 1940. Courtesy of the De Benedetti family.

Curie Institute in Paris before arriving in the United States and it was thought that his employ by the Dayton Project might give away its goal.

De Benedetti and the physicists at Bonebrake had two main areas of focus: the first was providing the means to measure the amount and purity of polonium produced by the chemists at each step in their extraction processes; the second was detection and measurement of radioactive contamination on both personnel and equipment. Beta and gamma radiation were measured using regular glass Geiger tubes with digital counters; from these measurements the amount of polonium present in the samples was calculated. Because alpha radiation cannot penetrate glass, standard Geiger tube technology could not be employed. As soon as he arrived in Dayton in 1943, Harvard-educated physicist John J. Sopka, 24, was sent north to Chicago's Met Lab to learn state-of-the-art alpha radiation measurement. He returned with two *a* (alpha) monitors that he adapted so that chemists could measure the alpha particle intensities of the polonium samples.[20] As the research grew more complex, so too did the physics. "This was a brand new field. Little instrumentation was commercially available, and there was little or nothing in the literature about alpha counting," Curtis recalled. The electronics department, under the direction of Josef Heyd, developed and built most of the instruments. "We developed measurement techniques which, when we were able to publish them, established us as authorities in the field," Curtis said.[21]

The physicists also undertook research and development of calorimetry and other instruments to measure small quantities of alpha-emitting radionuclides. De Benedetti, who was known for his creativity, developed a faster and more accurate calorimeter that was used in Dayton and at Site Y to accurately determine the half-life of polonium. "He never wanted to do the same thing twice. Once he'd worked out a technique, he never wanted to have anything to do with it again. He wanted something new and more challenging to work on," Curtis recalled of her supervisor.[22] The physicists also assisted in the mechanical development of micro-beam balances and electronic equipment, and undertook limited biological research on the toxic effects of polonium.[23] Counting and calorimetry were valuable not only in the laboratory, where scientists studied neutron sources and neutron-emitting impurities in purified polonium, but were also used for health safety monitoring.

In February 1945, health physics was further organized in Dayton when a clinical lab was set up in Unit III. More extensive survey and monitoring routines followed in April, when Lieutenant Bernard S. Wolf, MED senior medical officer, was loaned to the Dayton Project.[24] A letter to Thomas from Stafford Warren, medical director of the Manhattan Engineer District, detailed permissible levels of polonium following a meeting at Clinton on June 22, 1945, during which the topic was discussed. The maximum tolerance level was set at 3,000 disintegrations (1,500 counts) per minute per 24-hour output of urine.[25]

In May 1945, the Dayton Project expanded its facilities to include a six-story warehouse at 601 E. Third Street in an industrial section

10.3. The Warehouse (pictured in 2016 on the left) in downtown Dayton was leased for use as a health physics lab and equipment storage. Photo by author.

of downtown Dayton. Health physics laboratories were established in the building.

Until 1948, three floors of the building were used for equipment storage, a fourth floor served as office space, and a fifth floor was used as a health physics laboratory, where bioassay, environmental monitoring samples, and polonium's effects on laboratory animals could be studied and tested away from contamination by other polonium sources.

Due to polonium's tendency to migrate, shipments from site to site were known to be troublesome. Scientists at the receiving end of polonium shipments "learned to look for it embedded in the walls of shipping containers" when the foils came up short.[26] Los Alamos uranium and plutonium purification group leader Richard W. Dodson reported in December 1944 that the quantity of polonium received was as much as 22 percent lower than the quantity Dayton indicated it had shipped. "The deficit probably is caused by actual loss of the material from the foil during transit, and the missing material may be recoverable from the shipping containers," he wrote.[27] By January 1945, the problem was under better control. Of the two shipments from Dayton made that month, only 3.5% of the polonium migrated from the 12 foils and strips to the shipping container walls.[28]

Lyle Albright, a young chemical engineer from the University of Michigan, was drafted into the Army and sent as a member of the SED to work at Hanford canning uranium slugs and then in health physics. Albright handled the materials sent to Dayton but had no idea until many years later what he was shipping. One day, while monitoring the radioactivity of boxes at Hanford, he came upon a second health safety officer who was taking a break. As there were no chairs in the area, his co-worker was sitting atop a stack of boxes. Albright pulled out his Beckman instrument to measure radioactivity in the area and saw the reading rise as he approached the boxes. All the boxes were hot, he recalled. Neither he nor the fellow who was taking a break knew what the boxes contained, but they decided it was not a good idea to sit on them. The boxes were contaminated with polonium that had migrated out of the slugs. They were buried in the desert the next day.

"I didn't know anything about polonium or how the bombs were put together, but when we found that the boxes were contaminated, we immediately wanted to know more," Albright said. He got his answer in 2008, at the age of 87. "I guess we were filling them with slugs and sending them to Dayton, where they were purified," Albright said, astonished.[29] He had just learned about the connection between his work in Hanford and that in Dayton; Manhattan Project compartmentalization had succeeded.

CHAPTER 11

VE DAY AND DEADLINES

BY APRIL 1945, Hanford production reactors were fully operational.

"Yields of plutonium in the March runs in the 200-T Area (separation plant) continue to be very satisfactory," Seaborg reported.[1] With deadlines already stressing staff and a test date for the plutonium bomb set by the Project Trinity Committee in March for July 4, Oppenheimer and Thomas wrestled with delivery schedules and with the purity and quantities of polonium needed for the initiator.

In memo after memo, quantities and delivery dates were set and then changed, reset, and then updated once again. Initial plans called for monthly shipments of polonium from Dayton to Los Alamos beginning on March 1, 1945, and continuing through January 1, 1946. The amount of polonium was to increase in that time from an initial shipment of 10 curies per month to an ultimate delivery of 500 curies per month.[2] The schedules were, however, subject to frequent change, and Oppenheimer did not fully commit to using polonium-beryllium initiators until March 15, 1945. In June, the Los Alamos Chemistry and Metallurgy Division reported that "great progress has been made in the preliminary work necessary for the preparation of

11.1. Aerial view of the 100B Area at Hanford in January 1945. The B reactor is the square building at center right. U.S. Department of Energy, P-8015.

urchin initiators charged with *(censored)* Po and with a background neutron emission of less than *(censored)*. Many advances are being made in our knowledge of polonium, an element in some respects newer than plutonium." In the same report, Iral B. Johns, leader of the Los Alamos CM-15 (polonium) group, noted that "methods have been developed and tested and equipment is ready for preparation of the first complete initiator. This will contain *(censored)* quantity of Po and will be assembled June 2."[3]

Italian Fascist dictator Benito Mussolini was killed in Milan on April 28, 1945. On April 30, German Nazi leader Adolf Hitler committed suicide. And on May 7, in the midst of the heightening intensity of the bomb project, Germany surrendered to the Allies. The pressure to beat the Germans to the atomic bomb had been slowly lifting as intelligence had revealed that Germany was behind in its uranium research efforts, but war in the Pacific Theater with Japan now loomed as a great threat, and uncertainty about the Soviet Union remained.[4] The American bomb effort continued. On May 21, Oppenheimer requested that Monsanto provide "20 extra cups with 1/8 curie in the first half of July."[5] The following day, Thomas responded that a shipment of three cups was on its way: "This completes initial quote of 20. We now plan shipment of five cups each week with shipping dates on May 26, June 2, 9, 16, 23 and 30. By July 1, the quote of fifty will be completed."[6] On June 6, Oppenheimer confirmed the new direction.

> Dear Charlie:
>
> For your records, I am sending you a note about the polonium agreements which we reached last Friday. I hope they correspond with your understanding.
>
> I am also enclosing some very rough notes covering our discussions that evening. I wanted you to see them so that you may be sure I understood just what you had in mind.[7]

The schedule for Clinton-produced material allowed for five days of cooling and transportation, then 25 days in Dayton for the extraction and preparation of the polonium, followed by shipment to Los Alamos. On this timetable, bismuth would arrive in Dayton on May 1, and extracted polonium would ship to Los Alamos on June 1. The timetable for material produced at Hanford was longer, due to the geographic distance: cooling and transportation time from Hanford to Dayton was 10 days, plus 30 for extraction and preparation. Accordingly, bismuth discharged at Hanford on July 11 would be ready for shipment from Dayton to Los Alamos on September 1.[8]

The nearly daily communication between Los Alamos and Dayton about quantities and purity requirements indicates the quickly evolving research and development of the project, the immense importance of the polonium, and the pressure on scientists in Dayton and Los Alamos to deliver a bomb sooner rather than later. The critical material and process was closely guarded. Per dictate of Colonel Nichols, the materials in Dayton were from May 1945 forward to be referred to by code name—postum for polonium, case for curie, and soda pulp for bismuth. "Under this code, X cases of postum would mean X curies of polonium. Likewise, Y pounds of soda pulp would mean Y pounds of bismuth. It is not intended that the adoption of this code apply to your source of supply of bismuth. Also, it is not intended that the adoption of the code change the classification of correspondence," he wrote to Thomas on May 16.[9]

Preparations for an implosion test had begun as early as March 1944.[10] By June 1945, Dayton was shipping 35 curies of polonium to Los Alamos every week to meet the July 4 bomb test date. "The cups will be acceptable provided the amount of polonium lies between 0.1 and 0.15 curies per cup. The neutron background should be kept as low as possible, but 25 neutrons per cup per second will meet the specifications," Oppenheimer specified.[11] In July, as the bomb deadline neared, the Corps of Engineers issued a revised schedule for polonium shipments to Los Alamos using the code word "cases" for curies—deliveries were to occur weekly, beginning July 7 and running through August 25, 1945.[12]

July 14	35–40 cases
July 21	50–60 cases
July 23	50–60 cases
Starting August 4	75 cases per week
Starting December 1	125 cases per week

DELIVERING THE GOODS

Bismuth was delivered and polonium picked up for shipment by couriers who came and went from the Playhouse at all hours, purposefully keeping their schedules unpredictable. "Schedules were established for delivery of the purified polonium which were exceptionally hard to meet. It became an art to delay the courier arriving to pick up the polonium. Some deadlines were so close that an employee would be sent to talk with the courier and to keep him occupied while the final touches were put on the packages. Still, all commitments were met and shipments were made on schedule," wrote Keith Gilbert in his history of the Dayton Project.[13]

Max Gittler, who had studied mechanical engineering at New York University, was one of four GIs responsible for transporting the material from Oak Ridge to Santa Fe (the couriers did not enter Site Y), a 53-hour drive. They also traveled to the University of California and

Dayton, along with weekly runs to Chicago. In some cases, the material traveled by train; the first irradiated 25-milligram sample of Hanford plutonium was carried by courier from Site Y to Chicago aboard the California Limited.[14] Train couriers dressed in civilian clothing and carried the valuable cargo in standard luggage. They retreated to locked train compartments, and were eventually directed to wear film badges to measure the radioactivity exposure resulting from being sequestered with the materials.[15]

Gittler and his team drove a truck and car in convoy and knew that the material they were transporting in the back of the truck in a heavy lead pot gave off radioactivity, but they did not know what it was. They kept instruments in the car to measure the level of radiation they were receiving. When the radiation reached a maximum level, the pair of couriers in the truck rotated into the car, and those in the car moved over to the truck. "We were intrigued by the level of radiation that was increased as we traveled. I knew it was going down the periodic table, and we knew it was hot, thermally and in radiation terms," Gittler said.[16] The small but extremely heavy lead pot, which was crated and sat alone in the bed of the truck, gave the vehicle an unusual profile. "When we stopped for gas, the attendants would notice that the springs were almost fully compressed. It was pointed out to us so many times. It was unusual that a relatively small container sitting in the middle of the two and a half ton truck was bringing the springs down beyond what you would normally expect," Gittler said.[17]

The couriers were routed around all cities and stopped only to eat and get gas. Accidents were always a concern. A shipment of uranium scrap spontaneously ignited while on a train platform in Fort Wayne, Indiana, creating flames that shot 30 feet into the air and caused "local alarm."[18] Gittler described another mishap:

> We had a weekly run to the University of Chicago to Enrico Fermi, and one winter day we skidded. We were traveling in a van and the container was embedded in a crate, like a shipping crate. We skidded on the road. The crate flew out the back and skidded on the road. And fortunately, there was very little traffic. We were able to recover it, and the four of us were able to lift it up and get it on the bed of the van again and then put it back in place. It traveled quite a way on the

ice. The van whipped around, the back doors opened, and it flew out. There was no traffic. There were no bystanders. It was not an attraction to travelers.[19]

The couriers wore civilian clothes when they made deliveries to Dayton, though they were armed:

We carried guns, shoulder holsters, and after we left the compound we put on Tennessee plates. We found this to be a very high-class residential district, and we backed into the garage of one of the residences (the Playhouse) grouped in a circle. Inside was a laboratory. There was no furniture. Nobody lived there. It was an entire laboratory for polonium. . . . I cannot believe the adjoining houses were not aware of what was going on, they were so close and they would see a truck come up. Of course, it had Tennessee plates, it was a civilian truck, but nobody lived there. There was no contact with neighbors, but I am sure there were suspicions.[20]

Seaborg, who traveled frequently from Chicago's Met Lab to Site Y and Clinton, also acted as a courier. In July 1943, he and his wife vacationed in Santa Fe. They were sent before dawn to a restaurant to meet a man and collect a plutonium sample that the Met Lab had loaned to Site Y. "Bob returned the 200-microgram sample of plutonium-239, which he was guarding with his personal Winchester 32 deer-hunting rifle. . . . With the sample in my suitcase, Helen and I took a bus from Santa Fe to Lamy, where we boarded a morning Santa Fe train to Chicago," Seaborg recalled.[21]

CHAPTER 12

TESTING THE BOMB AT TRINITY

WHILE THEORETICAL and experimental work on the plutonium bomb
had progressed, one had not to that time been exploded. By the winter of 1944, Oppenheimer and Los Alamos division and group leaders
began to consider an atomic bomb test. The first formal arrangements
were made in March 1944 when the X-2 group (which became Project
TR) headed by Kenneth Bainbridge was created within Kistiakowsky's Explosives Division. By September, the Alamogordo Bombing
Base was chosen as the test site.[1]

On May 7, 1945, a 100-ton TNT shot with radioactive solution containing fission products from a Hanford slug was fired from a 20-foot
tower. This was a dress rehearsal for the Gadget test to provide calibration of blast and shock equipment. The ultimate test would use a
lens-implosion model in which explosive lenses would surround the
initiator to convert a multiple-point detonation and trigger implosion
shock waves. The first lens test shot had been made in November
1944, with constant improvements leading to a final lens design.[2] Lens
molds were to be ready for casting by the beginning of April 1945, but
an adequate supply of full-scale molds was not on hand until June.[3]

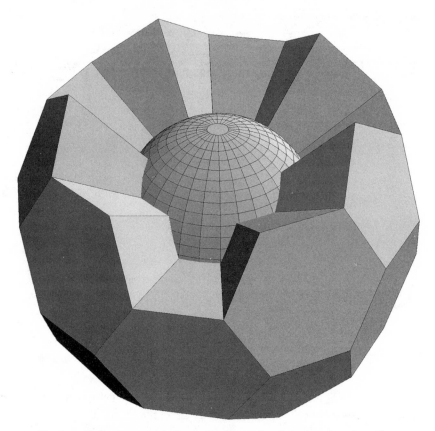

12.1. The lens-implosion model chosen for use in the plutonium bomb used explosive lenses and booster layers surrounding an initiator to convert a multiple-point detonation and trigger implosion shock waves. Thirty-two explosive blocks fit together in a pattern like a soccer ball. © Carey Sublette, "The Nuclear Weapons Archive," nuclearweaponarchive.org.

On June 14, Oppenheimer sent a memo to group leaders explaining that delays in fabricating the implosion lenses would push the test day later into July.[4] The new test date would not be before July 13 and would likely be July 23, Oppenheimer indicated. "In reaching this conclusion we are influenced by the fact that we are under great pressure, both internal and external, to carry out this test," he noted.[5]

The lenses, made by high-precision casting of explosive mixtures, were a crucial part of the explosive implosion system, which was made up of lens and booster layers. They surrounded the tamper and converted explosion waves into a convergent implosion.[6] The layers consisted of 32 explosive blocks (20 hexagonal and 12 pentagonal blocks) that fit together in the same pattern as a soccer ball. Each lens block had two components: the body made of high velocity explosive, and a parabolic low velocity explosive focusing element on the inner surface. These pieces formed the lens that shaped a convex, expanding shock wave into a convex converging one.[7]

Lenses weren't the only reason for the delay. All divisions were under pressure to meet the test deadline. In a June 30 review, division leaders reported that the earliest date their work would be complete was July 16. On July 5, Oppenheimer alerted Arthur Compton in Chicago and Ernest Lawrence in Berkeley to the test date: "Anytime after the 15th would be a good time for our fishing trip. Because we are not certain of the weather we may be delayed several days. As we do not have enough sleeping bags to go around, we ask you please not to bring any one with you."[8]

Final preparations for the test, which included preparation of the plutonium core at the abandoned George McDonald ranch house some two miles from the test site, ultimately began on July 12. A Teletype from Arthur Compton in Chicago to Oppenheimer on July 13 continued the fishing analogy and referred to Fermi (code name Farmer): "Can you let Farmer or Sam have a chance at the trout in my place. Best luck to catch the big one. Regret Chicago situation makes advisable stay here next week."[9] As with much of the science of the Project, success was an unknown. Thirty-eight hours before the test, a trial of the lenses indicated that they did not work, "which created enormous emotional outbursts about mine and everybody else's work," recalled Kistiakowsky. Despite this delay, the test occurred on July 16.

DAWN OF THE ATOMIC AGE

On Sunday, July 15, 1945, at 7 p.m., an Army sedan picked Thomas up in Santa Fe. Other visiting scientists were collected around 10 p.m. in Albuquerque, from which a caravan of three buses, three automobiles, and a truck carried some 90 people the 125 miles to the test site at the Alamogordo Bombing and Gunnery Range.[10]

In a letter written to his mother the day after the test, Thomas described the conversation as he and his travel companions rode to the test site. His car-mates in the Army Chevy sedan included William

Laurence, a *New York Times* science writer who had been designated by Groves as the official historian of the Project; Joseph Kennedy; and David Dow, executive assistant to Oppenheimer.

> Most of this conversation during the night ride was concerning statistics on the explosion about to be demonstrated, and our opinions oscillated back and forth as to whether the bomb would go off at all, or whether it would be a success. It was unanimously agreed that it would never be 100 percent efficient and our most sanguine hopes through the past three years were that the efficiency might lie between 5 percent and 10 percent.
>
> I noted that the trend of our conversation for the first three hours of riding was more optimistic and even small wagers were made that it would be somewhere between 5 and 20,000 ton equivalent to TNT. However, as this strange caravan drove on through the night, guarded very closely by GI MPs racing up and down the road on motorcycles, our optimism waned and around 1:00 o'clock in the morning we were feeling much more pessimistic. About this time the line stopped along the road where we had a rest and ate some sandwiches. At this time I had a chance to note the great number of seasoned troops which were in that area and a multitude of GI trucks and half-tracks [*sic*]. These soldiers were placed there not only to guard this large desolate area, but also this great fleet of trucks was called to evacuate people in the few small surrounding towns 60 to 150 miles away from the scene of the experiment.[11]

The convoy followed U.S. 85, then U.S. 380 to the Harriet Ranch turn-off and continued about 25 miles on a dirt road, arriving between 2 and 2:30 a.m. at Military Police Post No. 2, located about 20 miles northwest of zero point and 10 miles from Trinity Base Camp.[12] This was the primary VIP observation site, known as Compañia Hill.

Laurence reported that the group had an early morning picnic breakfast upon arrival and that it was chilly in the desert with a light drizzle falling. Some of the men tried to sleep on the ground. Thomas, Kennedy, and Laurence opted to sleep in the Army sedan, but the two chemists were kept awake by the reporter's snores. At 5 a.m., searchlights near the base camps started sweeping the sky, and the group in the car was roused by an officer who stuck his head in the car and

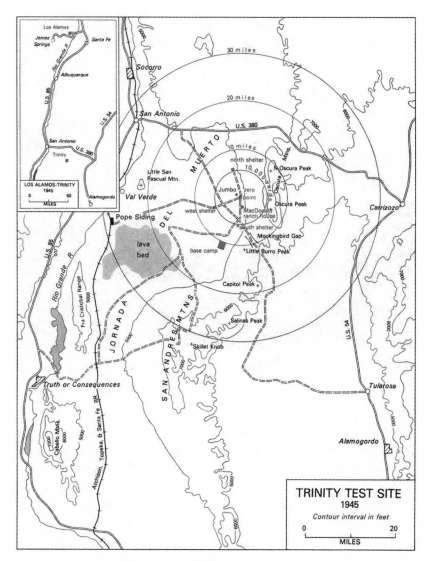

12.2. Trinity map. U.S. Department of Energy.

informed them the test would occur within 30 minutes. The group tried to make radio contact with "South-10,000" (the main bunker, located 10,000 yards to the south of Ground Zero), but their short-wave set didn't work. Physicist Richard Feynman—known for his electronics abilities—tried with no success to fix it, leaving only warning flares as the only signal that the test was about to occur.[13]

9.0 SEC.
N

⊢━━━┥ 100 METERS

12.3. Trinity at 9.0 seconds. Los Alamos National Laboratory. © Copyright 2011 Los Alamos National Security, LLC All rights reserved.

"The scientists stood around in the dark and munched candy bars as they waited for some divine revelation to tell them when the shot would go," Laurence reported. "The final countdown began at 5:10 a.m. with a crashing rendition of the 'Star Spangled Banner.' (KCBA in Delano, California, had crossed signals with the Trinity frequency and was opening its morning broadcast to Voice of America in Latin America." The scientists were told to lie flat or sit down on the ground and cover their eyes with pieces of dark welder's glass. The first man-made atomic explosion occurred at 5:29 a.m., with the blast reaching the group on Compañia Hill about 1.5 minutes after ignition. "An awed Charlie Thomas of the Monsanto Chemical Company shouted to physicist Ernest Lawrence that they had just seen the greatest single event in the history of mankind," journalist Lansing Lamont reported in his book *Day of Trinity*.[14] By 5:55 a.m., the observers had loaded back onto buses and left the site.[15] Kennedy later recounted the following as among the "high spots" of the Trinity trip:

"(Edward) Teller, (Hans) Bethe, and Thomas at about two minutes, applying sunburn lotion to their faces, with Teller saying '100 to 1 it's not needed, but what do we know?'"[16]

THOMAS AT TRINITY

Thomas's letter to his mother was written the day after the test, but perhaps because of security considerations was not mailed until August 29, 1945. He sent a similar one to his son, then a Navy enlistee, who carried it in his wallet until it was destroyed when the billfold got wet. He noted that he had written the letters "long before newspaper reports were released," but had waited until after the bombs were dropped to send it.

> At 5:30 a.m., July 16, 1945, I witnessed the first explosion of the atomic bomb. This date undoubtedly will go down in history for at that time man's world changed. It will take some time for the people of the world to know of this demonstration, and even after they know about it, to fully realize what it means.

A year later, Thomas shared his experience with a broader audience during his Commencement address at Washington University in St. Louis:

> As I lay there on the hot, moist sands of Alamogordo on the night of July 16, almost a year ago, I will never forget the tenseness of the moment. Here were about a hundred and fifty scientists who had been working, some almost four years, on the greatest cooperative scientific experiment that the world has ever seen. Midnight found it raining and the test had to be postponed. Then something was wrong with the electrical connection and the test was postponed again. Then finally around four o'clock in the morning, we knew definitely that the test would come off at five-thirty.
>
> At five minutes before five-thirty, we were gathered round in a semi-circle nineteen miles from the hundred foot steel tower on which the small atomic bomb was placed. Giant searchlights played upon the bomb from four directions, but from nineteen miles away down

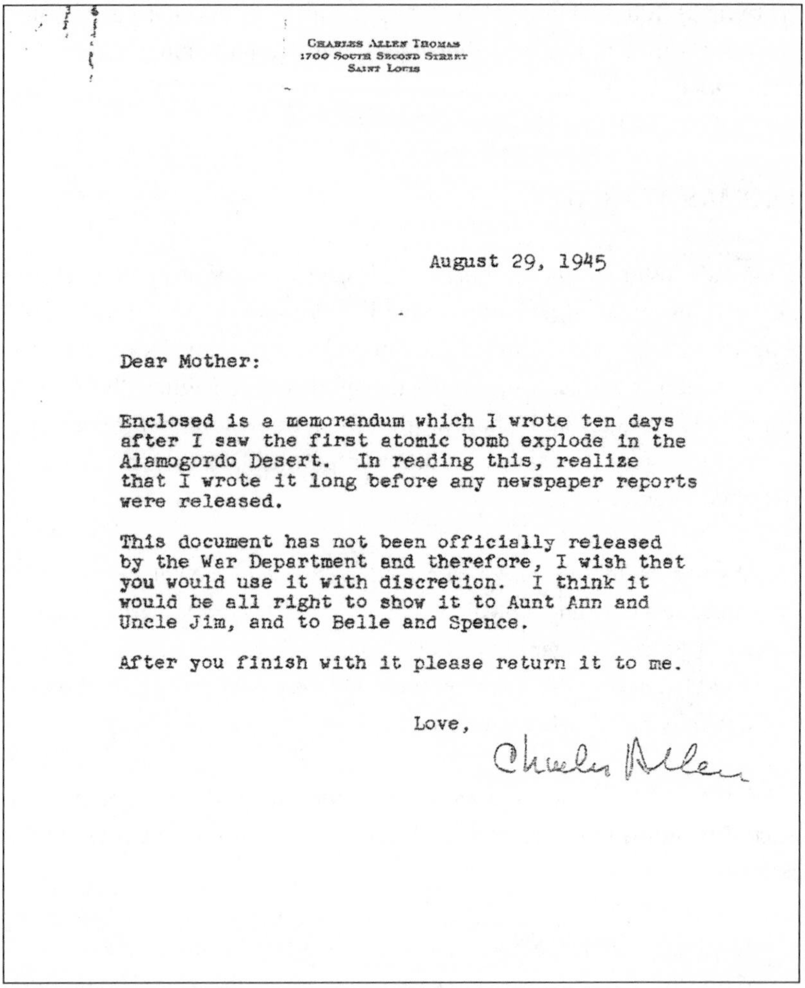

CHARLES ALLEN THOMAS
1700 SOUTH SECOND STREET
SAINT LOUIS

August 29, 1945

Dear Mother:

Enclosed is a memorandum which I wrote ten days
after I saw the first atomic bomb explode in the
Alamogordo Desert. In reading this, realize
that I wrote it long before any newspaper reports
were released.

This document has not been officially released
by the War Department and therefore, I wish that
you would use it with discretion. I think it
would be all right to show it to Aunt Ann and
Uncle Jim, and to Belle and Spence.

After you finish with it please return it to me.

Love,

Charles Allen

12.4. Trinity letter.

the shallow valley, we could only see a pinpoint of light. In this particular location, the Alamogordo Desert resembles a huge amphithe-
ater, the oval being about 40 miles on one axis and 60 miles on the
other. At the center of this oval was placed the atomic bomb. Some
nine or ten miles away was a small concrete fortress in which a few of
the crew who were to set off the atomic bomb were stationed. By our
radio equipment, we were in constant communication with this group
of men, and their nervous conversation only made the situation seem

more unreal. No one man was to set off the atomic bomb. There were some twenty-four different operations which had to be coordinated. It was particularly fitting that no one man threw the switch on the first atomic bomb, because it was the result of no one man, but of many men from all walks of life and from all nations on the globe, and here we were to see the demonstration of whether this great task which had been accelerated so much in the last two or three years was going to be a success. Up to this point, it had all been calculations, now we were to see whether these calculations were correct and whether it was possible to realize man's dream of the ages of transmuting matter into energy.

At 5:27, a Very light was thrown high into the air which told us that in three minutes the test would be set off. A few moments previous to that we had put on sunburn lotion, which seemed rather ridiculous in this very dark night, because we knew that even at nineteen miles, the radiation of infra-red and ultra-violet from this bomb might submit us to a very severe sunburn. We joked as we applied the lotion. We also were equipped with a very heavily tinted glass, such as a welder uses for electric welding. Here again, we knew that at nineteen miles the light intensity would be terrific. At 5:28 another Very light was thrown into the air and unfortunately this was a dud. Was this a bad omen? We shuddered as we thought what might happen, but the rocket thrown at 5:29 shot high into the air, its red light telling us that in one minute the greatest scientific experiment of the ages would be undertaken. The gravity of the situation weighed heavily upon us, and I will never know what went through the minds of my colleagues as we lay there on the sand that night. I could not but think of the cost and effort going into this experiment, something over two billions of dollars of money and years of concentrated effort on the part of thousands of American physicists, chemists and mathematicians. In a way this single experiment embraced the accumulation of two thousand years of research, beginning with the Greek scientist Democritus, who had the first conception of the atom. How significant it was that this experiment should be culminated in the world's great democracy and that it was a Greek, bearing the name of Democritus, who had the conception of the atom. Here was a free nation testing its scientific strength against a fettered nation; a race to learn whether a Fascist state and its philosophy would win against the philosophy of the free man and the Democratic state. Something like this must have been in the minds of all of us on that eventful night.

I kept the dark glass up to my eyes and shielded the sides with my cupped hands. I could see absolutely nothing and several times during these last few second before 5:30 a.m., I took the glass down to line up the tiny speck of light to be sure that I was looking in the right direction. Then, all of a sudden, I saw an intense pinpoint of light. This grew to a giant ball which rose rapidly in the air. It was awful. It looked like a giant mushroom, the stalk was thousands of tons of sand being sucked up by the explosion and the top of the mushroom looked like a flowering ball of fire. From the side of my glass and through my cupped hands, I could see the entire desert was brilliant. I turned completely around. It was intense daylight. The desert was completely illuminated. Later we found that the light intensity was equivalent to several noonday suns at nineteen miles from the experiment. Looking back through the glass, I found that the ball had grown tremendously. It was high in the air. The sand, which was the stalk of the mushroom, was now falling back to earth, carrying with it many curies of radioactivity. It was literally a sun coming up entirely too close. Then it began to dim and I took my glass away, but flinched even then with the terrific brightness of the light. It was still rising and growing in diameter, and as it faded the white light changed to orange and gradually into a purple cloud, which at times looked fluorescent. It resembled a giant brain, the convolutions of which were constantly changing in color. This effect was due to the ionization of the gases in this giant cloud which by now was several miles in diameter. Gradually, it faded and at approximately eight miles in the air, the cloud had reached the diameter of two or three miles. By this time, most of us were on our feet, completely forgetting that the shock wave was to follow, for as yet we had not heard a sound.

Then the shock wave came, a hot blast pressing against our bodies and a terrific clap of noise struck our eardrums as though we were the target of heavy lightning. This noise seemed to last for several seconds, and it rumbled through the desert, persisted and gradually faded away. The shock wave had reached us by my watch approximately a hundred seconds after the explosion. Ernest Lawrence (physicist and director of the Rad Lab at UC Berkeley) who was beside me embraced me and we both jumped up and down shouting, "It works!"

Everyone seemed to be dancing and shouting, slapping each other on the back and making inarticulate noises. The intense nervous tension was being released. The first rays of the natural sun could be seen coming up now in the East and the white cloud was now fanning out overhead, but it did not move as there was practically no wind. Ten minutes later, we reluctantly left the scene, and as I looked back through the window the giant cloud was now gray in color, still filling the horizon with an ominousness that was fear inspiring.

Yes, on July 16, 1945, man's world had changed, for he had created a weapon with which he might destroy himself and civilization, or he might prevent future wars with it.[17]

CHAPTER 13

HIROSHIMA, NAGASAKI, AND THE END OF THE WAR

PRE-ASSEMBLY PARTS for the uranium bomb, Little Boy, were shipped the day of the Trinity test to the Pacific Island of Tinian about 1,500 miles from mainland Japan. They arrived on July 26, with the target inserts following on July 28. A team of scientists and military specialists there assembled the bomb, which was ready for use by August 1. Delayed by a typhoon, Little Boy was dropped on Hiroshima, Japan, on August 6, 1945. An estimated 135,000 people were killed.

Three preassemblies for the plutonium bomb, Fat Man, arrived on Tinian on August 2, 1945. The bomb was dropped on Nagasaki, Japan, on August 9, 1945. An estimated 70,000 people died. A second plutonium core and initiator had been scheduled to ship to Tinian on August 12 or 13.[1] Japan surrendered on August 15, 1945.

President Truman's August 6, 1945, announcement to the American public regarding the use of the bomb on Hiroshima referred to the work of the Manhattan Project scientists: "The battle of the laboratories held fateful risks for us as well as the battles of the air, land, and sea, and we have now won the battle of the laboratories as we have won

13.1. Fat Man at Tinian Island, the staging base for delivery of the atomic bombs and one of the largest airbases of World War II. Los Alamos National Laboratory.

the other battles."[2] Dayton newspapers announced the bomb: "Allies Win Great Scientific Race, U.S. Drops New Atomic Bomb on Japs" and "Atomic Bomb, Most Destructive Force in History, Hits Japan."[3] Another article downplayed any Dayton connection: "The atomic bomb is news to military personnel at Wright and Patterson Fields, they said this morning. There has been no testing of the bomb here whatever, a survey showed. The Monsanto Chemical Company's Dayton plant also denied any knowledge of the bomb's development."[4]

Dayton Project laboratory director James Lum was in the garden of his Dayton home when news of the Hiroshima bomb came over the radio. He chatted with his neighbor, who marveled at the destruction. Lum did not disclose he had any part in it, though his role would be made public in newspaper headlines the next day.[5] At the Playhouse, Thomas gathered employees and revealed the nature of their work. The men and women of the Dayton Project, who had not known what they were building, finally learned, along with the rest of the world.

13.2. Betty Halley, 1948. Courtesy of Betty Halley Jones.

They were told not to share any information about their work. Mary Lou Curtis described learning about the bomb:[6]

> One day some of us were standing in the cafeteria line at Unit I when someone rushed in with a newspaper. The headline read, 'Atom Bomb Dropped on Hiroshima.' No phone calls were permitted in or out of the lab that afternoon, since our connection with the Manhattan Project was secret. When the second bomb was dropped on Nagasaki, and the war ended, we who had lived through the upheaval in the nation and in our personal lives following Pearl Harbor felt only intense relief.

Betty Halley, who had begun work in the Unit III counting laboratory only months earlier on May 5, learned with the others about the product of her work: "We were all called to the assembly room and a movie was shown of the dropping. One of the chemists fainted when we were shown the destruction the bomb had done."[7]

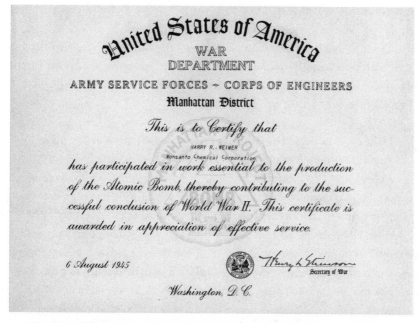

13.3. The War Department gave Dayton Project workers, including chemist Harry Weimer, certificates of recognition for their work on the atomic bomb. Harry Weimer Collection, Archives and Brethren Historical Collection, Funderburg Library, Manchester University, North Manchester, Indiana.

Headlines in Dayton newspapers the next day revealed Monsanto's role in the bomb effort: "2 Dayton Firms Help Produce Atomic Bomb."[8] The article also reported on contributions to the project by the local Duriron Company (450 N. Findlay Street) and its 350 workers. Duriron had provided corrosion-resistant equipment and alloy component parts for the atomic bomb to Manhattan Project facilities identified as in Tennessee, Washington State, the University of Chicago, and Berkeley.

RECOGNITION

The Manhattan Engineer District was quick to recognize those who had worked on the bomb, though it did not release details about the work. Dayton Project workers received certificates of appreciation from the War Department stating that they had "participated in work essential to the production of the Atomic Bomb thereby contributing to the successful conclusion of World War II."

Colonel Nichols wrote to Thomas on August 6, expressing his indebtedness for "the part played in making possible our hour of triumph."[9] Likewise, Groves communicated his gratitude to Monsanto President Edgar Queeny:

> I want you to know that Dr. C. A. Thomas and his associates made a major contribution to our success. Dr. Thomas personally coordinated a very important phase of the chemical research pertaining to the project; he also completed vital research and solved production problems of extreme complexity without which the atomic bomb could not have been. Not only did he successfully meet all the schedules originally set up, but he was able to multiply production many fold when requirements suddenly changed. We also called upon Dr. Thomas to give us aid and advice concerning difficulties we were having on many other parts of the project, and he always responded in an extraordinarily capable way. His cheerful personality, tactful manner and recognized ability made it possible for him to go into our plants or research establishments, study the situation, offer advice and have it accepted cheerfully and promptly by those in direct charge. Such wise and capable counsel was invaluable to our success.[10]

Groves also wrote to Thomas:

> You and your company played an essential role in the achievement of this huge success and early peace. I want you to know that I realize the work you did at Dayton, particularly your success in expanding the plant without adequate warning, made it possible for us to perfect and utilize the atomic bomb. Your efforts in coordinating the chemical research for the purification of plutonium and the many times you gave advice to me and others on the project were invaluable in assuring us that we were on the right course of action.[11]

Oppenheimer also wrote:

> I would like to add my own voice to theirs, because I am perhaps in a position to know more intimately how decisive were the contributions

that the workers of your company made and how helpless we should have been without their skill and devotion, and without your leadership . . . we could not have made the bombs without their help.[12]

Atomic Energy for Military Purposes, the officially sanctioned report on the atomic bomb project written by Princeton University physicist Henry DeWolf Smyth at the request of General Groves, was released to the public on August 12. The book, known as "the Smyth Report," was meant to inform the public about the project, paving the way for understanding. Any information other than what was described in the book was considered classified and not to be released. "The ultimate responsibility for our nation's policy rests on its citizens and they can discharge such responsibilities wisely only if they are informed. The average citizen cannot be expected to understand clearly how an atomic bomb is constructed or how it works but there is in this country a substantial group of engineers and scientists who can understand such things and who can explain the potentialities of atomic bombs to their fellow citizens," Smyth wrote in the preface.[13] The report contained the basic physics of the project, which was already generally known in the scientific community, but did not discuss the chemistry, metallurgy, or ordnance. The latter areas remained classified and, therefore, largely unrecognized by the public as having played a critical role in the project. The work in Dayton was declassified in 1983, with much of the chemistry not declassified until the late 1990s. Portions of the chemistry of plutonium, polonium, and implosion programs remain classified to this day.

Met Lab chemists noted the chemistry omission in the Smyth Report. "The Smyth Report, describing the American Atomic Bomb development has been issued," Seaborg reported on August 13, 1945. "Unfortunately, this does not adequately describe the contribution of chemistry to this extraordinary development."[14] Smyth acknowledged the omission: "In deciding which subdivision of the atomic-bomb project should be discussed first and most fully, we have been governed by criteria of general interest and of military security. Some developments of great technical importance are of little general interest; others both interesting and important must still be kept secret," he wrote.[15]

Monsanto employees who had worked on NDRC wartime contracts were given Certificates of Merit during a ceremony on August 17,

13.4. Dayton Project personnel, summer 1945. Mound Science and Energy Museum, Miamisburg, Ohio.

although those involved with the Dayton Project were not among the 188 names listed in the event program. The certificates, awarded by the U.S. government's Office of Scientific Research and Development, recognized "significant contributions to the war effort." L. H. Farinholt, who represented the NDRC at the ceremony, acknowledged that some details could not be shared. "An award is the thing next best to telling the full details of the work which still must remain partially secret, even from the workers," he said.[16] A newspaper account listed the accomplishments as including research and production relevant to the synthetic rubber program; its TNT program; its production and research with sulfa drugs, DDT and other helpful chemicals; the research concerned with the production of the atomic bomb . . . phosphorous production; plastic compounds and the treatment of leathers for GI shoes, and the company's aid to those in the textile field.[17]

Monsanto Chemical Company received an Army-Navy E Award, granted only to facilities that were considered particularly outstanding in production for the War and Navy Departments.

Harry Weimer's son, then 15, remembers his father's return home from the Dayton Project in August 1945: "We had been told he was working on a government project. . . . The only thing he said when he came home was, 'There's going to be some extraordinary news in the papers before too long. Keep your eyes peeled.'"[18] The chemist, who had finished out his Dayton assignment working in the Playhouse with production manager Ralph Meints, was conflicted about his work: "He came home and literally walked the floor all night long. Exhausted by morning he picked up a cup of coffee and remarked, 'We have succeeded, but I wish to God we hadn't. It's awful, but it has to be!'" his wife, Orpha Weimer, recalled.[19]

Betty Halley remained with the Dayton Project's successor, Mound Laboratory, until November 1979. She worked on detonator projects with related nuclear laboratories in Los Alamos, Livermore (Berkeley), and Albuquerque. Even 71 years after the conclusion of the bomb project, she admitted that she was still "very cautious what I talk about, for fear they may ask questions that I have strong reservations answering."[20]

In September 1945, Monsanto President Edgar Queeny shared news of the company's involvement in the bomb project with shareholders:

> The War Department has now granted us permission to advise you of Monsanto's contribution to the momentous advent of recent weeks—the atomic bomb. Until Japan felt the sting of nuclear energy, few if any Monsanto employees—other than those immediately concerned—surmised that the two hundred scientists and their associates, housed in Units 3 and 4 of our Central Research Laboratories, were engaged in this work which was organized under the conditions of secrecy characterizing the entire project.[21]

While he had already written to Thomas, General Groves wrote again in November 1945 to commend him for his work and, in a handwritten note attached to the formal letterhead, noted that he was doing so for the benefit of Thomas's children, and perhaps history: "Your assistance in the direction of certain phases of the work both at Los Alamos and elsewhere was an essential factor to our success. As you well know I leaned heavily on your scientific skill, judgment

13.5. Medal for Merit awards were given to recipients of the nation's highest civilian honor between 1942 and 1952. Thomas was one of several Manhattan Project scientists receiving the award—and the pins—"for extraordinarily meritorious conduct in the performance of outstanding services." Courtesy of the Thomas family.

and ingenuity."[22] In January 1946, Thomas was in a select group recognized for contributions to the war effort when he received a Medal for Merit from Secretary of War Robert P. Patterson.[23] Among other Manhattan Project personnel receiving the highest civilian honor were Oppenheimer, Fermi, and Kistiakowsky. The citation recognized Thomas for serving as "directing head of the production of vital substances required by the bomb; and he also coordinated the research work on an important phase of the project for the Manhattan Engineer District, Army Service Forces. A chemist of national distinction, Dr. Thomas' sound scientific judgment, his initiative and resourcefulness, and his unselfish and unswerving devotion to duty have contributed vitally to the success of the Atomic Bomb project."[24]

After the war, Dayton Project workers returned to their lives. Some of them resumed graduate studies, others returned to faculty positions or rejoined industry, and many of them joined other Manhattan Project scientists in becoming international scientific leaders.[25] They had been sworn not to discuss their bomb work, and many went to their graves with the secret. Implosion, which used polonium, was highly classified. Ralph Gates, a member of the SED who worked casting explosive lenses at Los Alamos, recalled his discharge at the end of the Project. "When we were being discharged, there was a discharge counseling, they gave us things we should never say. The word implosion, I remember, was one. You never let that word cross your lips. The implosion concept was totally secret."[26] Physicist John Sopka was told when he departed the project in August 1945, "Don't say anything, ever!" He first spoke of his experiences in 2003.[27] The wife of a Dayton Project veteran reported in 2006 that her husband never spoke of his work in the Project: "My knowledge of what was happening in Dayton is sparse to say the least, for after my marriage, I only met a few Monsanto people and my husband never discussed the Dayton years," wrote Orpha Weimer. Physicist Sergio De Benedetti's wife, likewise, said she knew nothing about what her husband had done in Dayton.[28] And so, with the silence of its participants, government records sealed, and the Dayton facilities largely destroyed, the full story of the Dayton Project has remained untold.

CHAPTER 14

POST-WAR

EVEN AS SCIENTISTS labored to build a bomb, their minds were on the post-war future. "The imagination of scientists on the Project, despite war time pressures, had already been sketching out broad programs and detailed uses for the postwar atom," wrote Alice Kimball Smith, wife of Los Alamos chemist Cyril Smith.[1]

From the Met Lab, Compton pushed for Groves to consider post-war plans for the research, expressing concern that the Manhattan Project's chemistry budget would be severely cut and that the scientists would disband and scatter with their accumulated knowledge as soon as the bomb work concluded. Early in 1944, Compton appointed a committee headed by physicist Zay Jeffries, a General Electric researcher and a Met Lab consultant, that was to consider the implications of the bomb and the field of atomic energy, along with issues of international control.

NUCLEONICS

The Jeffries Committee, one of many formed during the Manhattan Project to prepare for the future, included Thomas; Fermi; Hogness, director of the Met Lab's chemistry division; James Franck, associate director of the Met Lab's chemistry division; Robert S. Stone, associate director of the Met Lab and head of the health division; and Robert S. Mulliken, a physicist in charge of educational work and information director of the plutonium project. The committee produced a document titled "Prospectus of Nucleonics" in November 1944 that focused on the political as well as scientific issues of nuclear science and took for its title a name for the new field of nuclear chemistry and physics suggested by Jeffries, "nucleonics." The report stressed the importance to national health and security of the United States maintaining its lead in nuclear research and industrial application, a concern shared by many of the Project leaders. Jeffries advocated the development of the nuclear industry by private enterprise and suggested that an agency with both government and non-government representatives should be established to guide and coordinate nucleonics activities. For his part, Thomas supported research in peaceful applications of nuclear power that would involve collaboration between universities, private industry, and government. He also argued for an international organization to prevent nuclear armament. The report described a worldwide organization that would prevent the atom from "becoming the destroyer of nations."[2] Groves shared his own feelings about the future of nuclear research in a letter to Compton sent in February 1945:

> Since the basic consideration for any work performed under the direction of the Manhattan District must be winning the present war, it is necessary that the efforts of the District not be diverted in any way to post-war problems. The Military Policy Committee has concurred in these views. The Committee is also of the opinion it should not assume responsibility for the post-war period. I am in complete agreement with the recommendation that some commercial firm be found to take over the responsibility for the operation of the Clinton Laboratories. As you point out, they must be continued in operation for the production of vital materials and to carry on research development essential to the solution of recurring problems.[3]

Changes within the management and organization of the Project marked the approaching end of the military phase of nuclear science research and development. In early July 1945, Monsanto assumed control of Clinton Laboratories from the University of Chicago.[4] At the MED level, administration of the facility was transferred from the Chicago Area Engineer to the Office of the District Engineer in Oak Ridge. Monsanto's task as manager of the Tennessee facility was to continue production while exploring peacetime uses of nuclear fission, such as the production of radioisotopes, work on reactors using uranium, and the process of converting thorium to uranium-233. Thomas was appointed head of the new Monsanto division that controlled Oak Ridge laboratories, a position he held until the company gave up the Clinton contract to the Atomic Energy Commission in March 1948. In September 1945, he wrote to Groves expressing his concern about the U.S. legislature's apparent lack of interest in supporting nuclear research. Manhattan Project researchers were leaving Oak Ridge for positions that offered greater job security, he wrote. "The situation is very serious and unless something is done immediately, the splendid team of technical personnel at Clinton, and I'm told at other places, will disband with a resulting loss to the Nation."[5]

Within the MED, plans were made for ongoing government support of nuclear science research and development. During a meeting of Manhattan Project leaders at the Madison Square Area offices of the MED on February 9, 1946, an Advisory Committee on Research and Development was formed that included Thomas, Los Alamos physics division leader Robert F. Bacher, Arthur Compton of the Met Lab, chemical engineer and Manhattan Project advisor Warren K. Lewis, science advisor Richard Tolman, and leading Hanford physicist John A. Wheeler. In March, the committee recommended the establishment of national laboratories to supplement the work of District-supported research and development at private and university facilities. Met Lab activities, which had been under contract with the MED, were subsequently transferred to the University of Chicago's Argonne site on June 30, 1946.[6] Argonne National Laboratory was to be operated by the University of Chicago and a consortium of Midwest universities. Brookhaven National Laboratory in New York would, likewise, be formed by a consortium of northeastern universities and would also take national laboratory status. Both would be run by the MED or its successor, the Atomic Energy Commission. The advisory committee felt strongly that the government, in tandem with local universities and research institutions, should support national regional laboratories undertaking research and development in nuclear science.[7] As an industrial leader, Thomas was particularly interested in using nuclear power to generate electricity and the possible industrial applications of atomic energy. In April 1946, Monsanto accepted responsibility for the design, construction, and operation of a high temperature pile for power generation at Clinton known as the Daniels Pile. The plant was never built.

THE ATOMIC ENERGY COMMISSION AND MOUND LABORATORY

The Manhattan Engineer District came to an end with the Atomic Energy Act (the McMahon Act) on August 1, 1946. The act formed the Atomic Energy Commission (AEC), which assumed management of nuclear research, manufacturing, and testing sites on January 1, 1947. The Armed Forces Special Weapons Project, a joint Army-Navy organization, took over MED military responsibilities, while Monsanto assumed Site Y's urchin production: "Our problems now lie in maintaining the 'know-how' on urchins, and occasional technical developments and improvements, rather than in the production of combat units," Oppenheimer had written to Thomas in September 1945. He specified that Dayton should maintain a polonium production capacity of "10 units a month."[8]

With Dayton's Manhattan Project polonium facilities (Units III and IV) set to close and Site Y terminating initiator production, the need for a new polonium research and production facility was met by the creation of the Dayton Engineer Works and the construction, beginning in 1946, of a new facility near Miamisburg, 10 miles south of Dayton.

Known initially as Dayton's Unit V, the nation's first permanent Atomic Energy Commission facility was designed and operated by Monsanto. The facility was named Mound Laboratory, after a nearby Indian mound, and was located on 180 acres above the Miami River. It included 14 major buildings with a total floor area of 366,000 square feet, among them the Technical (T) Building, a vast underground facility designed to withstand torpedo attack. T Building was built to be operated for a month by 200 people without any supply or service

14.1. Mound Laboratories, c. 1949. Mound Science and Energy Museum, Miamisburg, Ohio.

from the outside.[9] Mound was tasked with continuing Dayton's Manhattan Project work on polonium-beryllium initiators. The facility was occupied in May 1948, with polonium processing beginning in February 1949. The opening of the Miamisburg facility was celebrated with the "The Atomic Energy Show: Living with Atomic Energy," a three-day community event sponsored by the Dayton Council on World Affairs, Monsanto, Miamisburg Civic Association, and the Atomic Energy Commission. The event, which took place April 29 to May 1, 1948, in the auditorium and gymnasium of Miamisburg High School, was said to be the nation's first public presentation of fission and chain reaction concepts, and featured public lectures and exhibits, as well as programs for students from 63 area high schools. A newspaper account described the show as dispensing "simple facts, simple information, with the hope that after the facts are presented, we can think them over and decide what role atomic energy should truly play in this world of ours."[10]

14.2. High school students at the Atomic Energy Show in Miamisburg, Ohio, April 1948. MS-484, the Miami Valley Atomic Energy Show Collection, Special Collections and Archives, University Libraries, Wright State University, Dayton, Ohio.

Attended by nearly 7,000 people, festivities opened with a keynote address by Arthur Compton, who had led the University of Chicago's Met Lab, on "The Moral Implications of Atomic Energy." It also featured exhibits including an interactive nuclear power plant that participants operated by pushing a rod into a pile, a ping-pong pile that simulated fission, and lectures by Mound scientists that were designed to "demonstrate living with Atom Energy and how the dreadful weapon of destruction is being turned to benefit industry, medicine, agriculture and mankind."[11] High school students attended matinees, received atomic energy comic books, and took part in an essay contest, writing on the topic of "How Atomic Energy Will Change Dayton 50 Years from Now."[12] "Scientists are turning into politicians these days and politicians are becoming scientists in an attempt to understand and control this bear we have by the tail," Malcolm Haring, director of Mound Laboratory, told audiences. "We hope through these talks and illustrations to give the man in the street the history and behavior of this bear."[13]

The Mound site was expanded over the years to include integrated research, development, and production supporting weapons, energy, and space missions. At its peak, the facility covered 305 acres with 116 buildings and a payroll of more than 2,500 workers.[14] In the early 1950s, technicians there processed polonium for use in power generators in the nation's first satellites. Initiator production continued at Mound Laboratory through mid-1969, with neutron source production and commercial sales of polonium continuing through 1972. Polonium-210 work ended in 1975.[15] U.S. Congress abolished the AEC in 1975 and incorporated nuclear weapons production into the Energy Research and Development Administration, which in turn became part of the Department of Energy in 1977. The Mound site was decommissioned in 1993, and the area remediated. It is now home to the Mound Advanced Technology Center. Its history is preserved in the nearby Mound Science and Energy Museum.

THE PLAYHOUSE COMES DOWN

All work at the Playhouse ceased on December 23, 1948, when polonium processing was transferred to Mound Laboratories. Although the Army Corps of Engineers had made an agreement with the Talbott family to return the building after the war, the interior of the facility was contaminated with radiation beyond repair.

The Playhouse was dismantled in February 1950, loaded into 55-gallon drums, and carted away in some 100 truckloads to the Mound facility or sent to Oak Ridge.

In addition to the buildings, cobblestones from the driveway and soil to a depth of seven feet under the foundation were also removed from the site. The AEC paid the Talbott Realty Corporation $138,750 for the structure and returned the remediated property to the family.[16] All that remains of the Playhouse is one brass doorknob and a small portion of the greenhouse roof, now in the possession of the Mound Science and Energy Museum. The Oakwood neighborhood stands much as it did at the time, with a new cul-de-sac adjacent to the Playhouse site and a large private residence occupying much of the Playhouse footprint.

14.3. One of the Playhouse greenhouses at the conclusion of the Dayton Project. The greenhouses were used as loading docks. National Archives (Atlanta), Records of the Atomic Commission.

The former Bonebrake Seminary, Unit III, had acceptable levels of radiation, so it was decontaminated, equipment was removed, and the building was returned to the Dayton Board of Education in 1950, which condemned it and tore it down later that year. All that remains of Unit III are a few abandoned single-story concrete block buildings in a run-down neighborhood. Central Research Department head-quarters, Unit I, on Nicholas Road was torn down in 1988. The 19-acre property was purchased by Quality Chemicals, Inc. in 1992 and sold to DuPont in November 2002.

After the war, Thomas rose quickly in the ranks of Monsanto, and the family relocated from Dayton to the St. Louis suburb of Ladue. He was elected vice president and technical director of Monsanto in 1945 and was named a member of the company's executive commit-tee that September. Thomas was appointed executive vice president and member of Monsanto's finance committee in 1947, was elected vice chairman of the executive committee in September 1948, and

14.4. Barrels of radioactive material fill the Playhouse floor as clean-up begins. National Archives (Atlanta), Records of the Atomic Commission.

14.5. Contaminated material is trucked out of the Playhouse. National Archives (Atlanta), Records of the Atomic Commission.

14.6. The skeleton of the Playhouse stands sentry over the Oakwood neighborhood. National Archives (Atlanta), Records of the Atomic Commission.

14.7. The Playhouse site, with all structures gone, was remediated and returned to the Talbott family in 1950. National Archives (Atlanta), Records of the Atomic Commission.

then named chairman a year later. He became president of Monsanto in 1951. Within his discipline, he served as chairman of the American Chemical Society, was named to the National Academy of Sciences, and was a founding member of the National Academy of Engineering.

ACHESON-LILIENTHAL COMMITTEE

In a letter to Thomas written November 1, 1945, General Groves expressed the opinion held by many that dropping the bomb on Japan had prevented the loss of lives of thousands of American soldiers in a Pacific Theater battle:

> All thinking persons must agree that the major final factor which determined the surrender of Japan was the atomic bomb. That surrender was an ultimate certainty without the bomb, but the war would have continued for weeks and most likely months longer had it not been for the use of our weapon. Many thousands of Americans were saved from death and tens of thousands more from hardship and injury.[17]

While the war with the Axis Powers and Japan had ended, the threat of the Soviet Union or other nations having a nuclear bomb had not been resolved and remained a concern. Shortly after the Japanese surrender, the U.S. Central Intelligence Agency estimated that the Soviets would not have an atomic bomb until 1950–53 and some said 1970. In January 1946, Undersecretary of State Dean Acheson appointed a five-man Committee on Atomic Energy chaired by David Lilienthal, head of the Tennessee Valley Authority, to study international inspection and the nuclear potential of various nations. The committee work was to inform government opinion on worldwide atomic control when the anticipated United Nations (U.N.) Atomic Energy Commission was formed. Thomas was named as a consultant to the Acheson-Lilienthal Committee, along with Oppenheimer; Chester Barnard, president of New Jersey Bell Telephone Company; and Harry Winne, vice president of General Electric Corporation. Oppenheimer and Thomas were the only committee members who had technical knowledge of nuclear energy.

14.8. Members of the Board of Consultants to the Secretary of State's Committee on Atomic Energy enjoy a rare moment of relaxation at Oak Ridge's Guest House (Alexander Inn) during their 11-week marathon beginning January 1946 to prepare "A Report on the International Control of Atomic Energy." The document informed the U.S. position on international control of atomic energy. From left: J. Robert Oppenheimer; Charles Allen Thomas; Herbert Marks, assistant to committee chairman and Under-Secretary of State Dean Acheson; and Carroll Wilson, assistant to Vannevar Bush. Photograph by Ed Westcott.

The committee met for the first time on January 23, 1946, in an oak-paneled room on the top floor of the American Trucking Associations building on Sixteenth Street in Washington, DC. The office had previously served as OSRD headquarters. Daniel Lang of *The New Yorker* provided a close look at the committee's work:

> The large room was drab. Each man had a desk or table and a kitchen chair. Telephones stood on the floor and the windowsills. There for two days amid the cobwebs, Oppenheimer put his colleagues through a short course in nuclear physics. Except for Thomas, it was the first time the panelists had been exposed to the physicist's extraordinarily fluent, lucid speech.[18]

In 11 weeks of intensive work, the committee produced "A Report on the International Control of Atomic Energy," informally known as the Acheson-Lilienthal Report. It was the first plan to propose international control of atomic energy via a proposed international Atomic Development Authority (ADA), which would monitor and develop nuclear power. "The consultants talked atomic energy in offices, in Pullman compartments, and aloft in an Army plane. Sometimes they deliberated for as long as eighteen hours in a day. They ate and slept and wrangled late at night in places that weren't home to any of them, and then, as soon as they rose in the morning, they would meet again at the breakfast table and resume their marathon discussion," Lang wrote.[19]

Herbert S. Marks, assistant to Acheson, described the intensive work: "Discussions and lectures, lectures and discussions. . . . We'd send out for sandwiches, and we smoked too much. The conference room got to be quite a mess, but none of the men in it would have been aware of it if it hadn't been for a cleaning woman who persistently tried to get in and was just as persistently shooed away." The discussions would last for hours and continued during site visits across the country.[20]

Thomas was a proponent of nuclear energy, which he viewed as an inexpensive and safe form of power and a means of producing radioisotopes for use in medicine and industry. "We will only reach our stride when suitable world controls over destructive phases of this new science are in operation. The alternative may be an atomic armaments race which will throw beneficial developments back many years or, if atomic war results, perhaps halt them forever," he stated.[21] He suggested formation of an international corporation to control uranium and thorium and their mining and primary processing with shareholders being participating nations. The approach complemented Oppenheimer's suggestion of an international control agency with monopoly of raw materials.[22] Marks, assistant to Acheson and an *ad hoc* member of the committee, later remarked that Thomas' proposal to internationalize uranium and thorium mines was the most critical and controversial feature of the plan. "It was Thomas' way of getting to the heart of the problem, of defining the limits. It was not a limitless problem, he said. The report stated: 'Uranium is the only natural substance that can maintain a chain reaction. It is the key to all foreseeable applications of atomic energy.'"[23]

Lilienthal gave each committee member responsibility for drafting a section of the report; Lilienthal, himself, would write the introduction. Oppenheimer was responsible for a "primer" on atomic energy; Wilson would cover raw materials, separating isotopes, plutonium production; Winne took the section on control systems; Barnard would discuss techniques of control, accounting and inspection, denaturing, and free association among scientists; and Oppenheimer and Thomas, individually, would cover the topic of a world authority.

> Thomas cast doubt on the adequacy of inspection alone. No nation would enjoy having a small army of inspectors descend on its laboratories and factories. ... Thomas thought it more practical to invest ownership of the world supply of uranium and thorium in an international commission, a cartel, or a world corporation. This agency would refine the metal and lease it to reputable nations or individuals for peaceful purposes under careful accounting procedures. This would not eliminate inspection, but theoretically, at least, it would simplify it. Any breakdown would be a sign of gathering war clouds.[24]

The committee's "Report on the International Control of Atomic Energy," which consisted of four volumes, was presented to the State Department on March 16, 1946. Each committee member was given a section of the first volume to read aloud. Oppenheimer, Winne, and Barnard presented Section II, "Principal Considerations in Developing a System of Safeguards." Oppenheimer read first. Thomas read the last chapter of Section II, related to how to draw the line between dangerous (essential to making an atomic weapon) and safe (radioactive isotopes for scientific, medical, and technological research and small nuclear reactors for radiation sources) activities. With revisions made, a final version of the report was released by the Department of State on March 28, 1946. It was the basis of the Baruch Report on international arms control presented by American delegate Bernard Baruch to the U.N. Atomic Energy Commission on June 14, 1946.[25]

The primary message of the Acheson-Lilienthal Report was that control of atomic energy through inspections and policing operations was unlikely to succeed. Instead, the report proposed that all fissile material, mining operations, and licensing of nuclear materials for peaceful research use be controlled by an international body known as the Atomic Development Authority (ADA). "Paul H. Appleby of the Bureau of the Budget was so excited the night he read it he could hardly sleep. 'In my opinion,' he wrote to Acheson, 'it is the most important most perfect governmental job that has been done in generations. Anyone who had anything to do with it can feel that his life has an extraordinary and enduring significance.'"[26]

The report proposed that the United States abandon its monopoly on atomic weapons and reveal what it knew to the Soviet Union, in exchange for a mutual agreement against the development of additional atomic bombs. This concession to the Soviet Union, along with how to handle violations, was one of the chief points of disagreement among the many interests concerned with drafting the final plan, a group that included President Truman, Baruch, military leaders, and the committee. The Acheson-Lilienthal Committee, for example, proposed issuing a warning to those in violation. Baruch's version called for the ADA to oversee the development and use of atomic energy, manage any nuclear installation with the ability to produce nuclear weapons, and inspect any nuclear facility conducting research for peaceful purposes. The plan also prohibited the illegal possession of an atomic bomb, the seizure of facilities administered by the ADA, and punished violators who interfered with inspections.[27]

Baruch felt strongly that in order to assure peace, the plan should contain more than a self-policed desire to get along; it needed clear sanctions for violations, though he did not define the punishments. In his version of the plan, the ADA would answer only to the U.N. Security Council, which was charged with punishing those nations that violated the terms of the plan by imposing sanctions. The Baruch version, however, stripped the U.N. Security Council of any veto power concerning sanctions. This, combined with U. S. continued insistence on retaining the bomb until it was satisfied with the effectiveness of international control, did not meet with the Soviet Union's approval and ultimately led to the plan's failure.

Thomas, who delivered the Commencement address at Washington University in St. Louis the day before the report was presented to the U.N., foreshadowed its contents in his remarks to the graduates: "Without a firm control over atomic energy, the United Nations will fail. It must be made abundantly clear to all people of all nations that this common danger now exists among all nations. And with this common danger must come the desire to seek togetherness to save all from destruction. . . . The individual must learn that his very safety depends upon nations working together."[28]

The plan needed unanimous approval to pass, but at the final vote on December 30, 1946, only 10 of the 12 member countries were in favor; the Soviet Union and Poland abstained. Several years later, the Soviet Union's reluctance to participate was explained. With the help of Soviet spy George Koval, who was embedded in the Manhattan Project and active in Dayton, the Soviets had been working on their own atomic bomb. They entered the nuclear arms race with the detonation of their first atomic bomb on August 29, 1949. The U.S. government had anticipated but misjudged the timing of the event.

Thomas believed the committee's work had merit: "What happened in 1946 was that a group of very different individuals was entrusted to think through a very difficult problem. Our final plan did not have all the answers. But it was a good starting point, and I believe today that it would have worked," Thomas said of the Acheson-Lilienthal Committee's work.[29]

Thomas continued throughout the rest of his life to push for industrial use of atomic energy, hoping first that Monsanto could develop the Daniels reactor as a power-demonstration plant at Clinton Laboratories. Thomas proposed an industrial study of nuclear power:

> Thomas suggested that industry be allowed to design, construct, and operate atomic power plants at its own expense, to produce both useful power and plutonium. . . . A dual-purpose reactor would give the Commission an additional source of plutonium at the very time it was endeavoring to increase plutonium production for weapon requirements. If Thomas could entice the Commission to accept such an agreement, private industry would have a compelling reason for access to classified technical information. Furthermore, revenues from the sale of plutonium to the government could be used to offset power costs and make the dual-purpose reactor more attractive to electric power companies. Thomas thought this incentive, plus the promise of long-term amortization, would induce private industry to undertake the huge capital investment required.
>
> Thomas's proposal was sufficiently attractive to command extensive study by the Commission's staff in the summer of 1950.[30]

14.9. Manhattan Project leaders gathered in 1946 to honor Arthur H. Compton (physicist and director of the Manhattan Project's Metallurgical Laboratory).

Standing from left: Charles A. Thomas, James B. Conant (chemist, president of Harvard University, Manhattan Project science advisor, and director of the National Research Development Council), Arthur H. Compton, Eger V. Murphree (Standard Oil of New Jersey executive and member of the S-1 Uranium Committee), Crawford Greenewalt (chemical engineer and liaison between Manhattan Project scientists and DuPont's work at Hanford).

Seated from left: General Leslie R. Groves (director of the Manhattan Project), Vannevar Bush (engineer, director of the U.S. Office of Scientific Research and Development, and Manhattan Project science advisor), Enrico Fermi (physicist), Colonel Kenneth D. Nichols (district engineer, U.S. Army Corps of Engineer, Manhattan Engineer District), George B. Pegram (Columbia University physicist and member of the S-1 Uranium Committee), Lyman J. Briggs (physicist, director of the National Bureau of Standards, and chairman of the Uranium Committee). Courtesy of the Thomas family.

An article in the April 15, 1946, issue of the *Bulletin of the Atomic Scientists* announced plans for the Clinton power plant:

> An experimental power plant is to be built at the Clinton Engineer Works at Oak Ridge, Tennessee. In broad and general terms the new pile will employ scientific principles utilized in the Chicago Pile where the first chain reaction in history took place, the Clinton Laboratories Pile where the first experimental quantities of plutonium were made, and the full-scale Hanford Pile in Washington, where plutonium is produced. But unlike the others, which were made specifically to produce plutonium, the atomic power plant will have as its primary purpose, the generation of heat energy and the conversion of that energy into electricity.

Thomas continued as a consultant to the government. In 1950, President Harry Truman appointed him chairman of the Scientific Manpower Advisory Committee of the National Security Resources Board to help coordinate government and private scientific research and planning for defense. He was also named chairman of an 11-man committee to advise Defense Mobilization Director Charles E. Wilson on classified research.[31] He believed in the power of cooperation between academia, industry, and government, stating, "The chemical industry can best serve mankind through a fusion of fundamental and applied research, implemented by the cooperation, understanding and mutual respect of all scientists everywhere."[32]

In 1957, Thomas was a member of a group that advised Secretary of Defense Neil McElroy to create an office to undertake advanced research projects that would keep the United States Defense Department technologically competitive. The resulting Advanced Research Projects Agency was founded in 1958 and became the Defense Advanced Research Projects Agency in 1972. According to his son, Thomas had by the late 1950s grown weary of government work:

> He developed doubts about the enlarging federal government and the scientists who remained in government service. By the time the Oppenheimer security case came along, he had been off the scene for some years. While admiring intelligence, he developed severe doubts

about intellectuals—both real and fancied—and their attempts to influence political events.[33]

Though his direct work with nuclear science was done, Thomas remained watchful and outspoken about the nation's nuclear science and weapons policies. In a 1961 address to the Houston Rotary Club, Thomas spoke out against the government's ban on nuclear testing and revealed continuing suspicions of the Soviet Union's nuclear program. He reminded his audience that the United States had been naively puzzled by the Soviet Union's reluctance to accept the proposed post-war plan for an international body to control nuclear materials and research, only to see the first Soviet atomic bomb test a few years later. He was especially critical of the U.S. government's decision in 1958 to cut nuclear testing and saw this as severely weakening homeland security. "I want to be completely understood," he told his audience. "We must not have another world war. But I believe history shows us that the correct way of preventing war is to be strong. Hoodlums don't try to open a well-guarded bank vault. Peace is earned by moving decisively from a base of strength, and not by shifting feet continually in an attitude of weakness."[34]

Far into life, Thomas continued to argue for the peaceful use of atomic energy and remained a staunch advocate for the nation's scientific strength throughout the remainder of his life, continually warning that the U.S. could not afford to lose strength in its engineering, scientific, and technical areas. Government support of education and research, he advised, was key to the nation's competitive edge and to its national security. He was described as "an articulate spokesman for basic research and higher education,"[35] and during his presidency of the chemical company grew Monsanto's research investment from $6.2 million to $101.4 million.

At the age of 80, two years before his death, he chided the government for ignoring the potential of alternative energy sources including nuclear-powered electric generating plants:

A modern industrial society depends on the continuance of a plentiful supply of energy.... It is time for America to wake up.... Because this country has failed to draft and follow a sensible energy policy, Americans now find themselves paying out $60 billion a year for imported oil. Yet virtually nothing has been done to speed up the production of coal and synthetic fuels or tap our huge oil-shale reserves.[36]

Though a proponent of nuclear power, Thomas had long had doubts about man's ability to handle the complex moral, ethical, and technical issues facing scientists. Speaking in 1954, he had expressed these doubts:

> Considering the last fifty years, what fantastic changes science has wrought in all our frames of reference, what shocks have been imposed on the human consciousness. . . . Has mankind advanced sufficiently in his moral habits to be trusted with the concepts, the extension of the senses, the machinery that science is putting at his disposal? Have ethical systems been enlarged, refined—modernized, if you will, to accommodate the complex, new relationships between individuals, groups, and nations made possible by developments in science? Has man's spirituality deepened, taken strong enough roots in his being, built a firm enough foundation so that the multitude of gadgets and material things will not crowd spiritual aspirations out of his consciousness? . . . The discordance between what science has shown us how to make and how we employ these technologies has some distressing results. . . . [37]

Thomas died on March 29, 1982, at his pecan and peanut farm in southwest Georgia, and is buried in Lexington, Kentucky. He was 82. He spent his final years focused on his lifelong interest in farming, trying to grow the perfect pecan. Many of the nuts he produced were named after his favorite songs. His competitive spirit lived on until the end, whether gathering his grandchildren around a poker table and showing off his card skills or continuing to hunt quail from horseback at an age when many wouldn't consider mounting a horse. Marnie had died in 1975, and Thomas was married to Margaret Chandler Porter in 1980. She survived him, as did his four children and eight grandchildren. He never spoke to the family about his Manhattan Project work.

APPENDIX I

SCIENCE PRIMER

URANIUM

By the time the Manhattan Project organized, uranium-235 was known to be fissionable, so it was the first route to the bomb selected. Uranium occurs in nature, but only 0.7 percent of the naturally occurring ore is fissionable U-235.[1] Uranium compounds extracted from ores were converted into the uranium oxide metal needed for the bomb by Mallinckrodt Chemical Works in St. Louis, Missouri, and then by processes developed at Iowa State College in Ames under a group led by chemist Frank Spedding.[2] The Ames laboratory was tasked with developing large-scale production methods for making pure uranium and a procedure for casting the metal. The uranium was ultimately enriched in the Y-12 electromagnetic separation plant at Clinton Engineer Works in Tennessee (now known as Oak Ridge). There was just enough enriched uranium for one bomb—Little Boy, which employed a simple gun-type design. Because the uranium bomb was inefficient—it contained 64 kilograms (141 pounds) of enriched uranium, less than a kilogram of which underwent nuclear fission—scientists also pursued a plutonium bomb.

PLUTONIUM

Plutonium, used as the fissile material in the Fat Man bomb, is a man-made element. Plutonium (Pu) was discovered on December 14, 1940, by chemists Glenn T. Seaborg, Arthur Wahl, and Joseph Kennedy at the University of California, Berkeley. It was produced by deuteron bombardment of a uranium-238 isotope in the university's 60-inch cyclotron. This produced a new element, 93 (neptunium), which decayed to another isotope, element 94 (plutonium).[3] Joining them in the research that winter was Emilio Segrè, a physicist who had emigrated from the University of Palermo. In the spring of 1941, Seaborg, Segrè, and Kennedy discovered the plutonium-239 isotope and showed that it would fission under neutron bombardment and could be used in a similar way to uranium-235 in gun-assembly bombs.[4] In Seaborg's words: "Plutonium (element 94–239) is produced by allowing a uranium-238 atom to absorb a neutron, and then emit two beta particles.[5] It is produced by bathing uranium salts in the neutrons from cyclotrons, followed by chemical extraction."[6]

To be considered fissile, an isotope's atomic nucleus must be able to break apart (fission) when struck by a slow-moving neutron and to release enough additional neutrons to sustain the nuclear chain reaction by splitting further fissile nuclei. One of the isotopes produced by neutron bombardment of uranium-238 in a reactor is plutonium-239. Extracting plutonium-239 from the spent reactor fuel on a large scale was one of Manhattan Project chemistry's major accomplishments.

It is proposed to make 49 (plutonium-239) in much larger quantities by producing a fission reaction with uranium that will be self-sustaining," Seaborg wrote. "When the fission of a uranium atom occurs, two atoms of roughly half its mass split apart with high energy (the fission products), and various radiations are emitted. These radiations include neutrons . . . one of the neutrons will be caught by a second U-235 atom to produce the next fission stages, while the remaining neutrons are caught by the heavier U [uranium-238], which in time changes to 49.[7]

As fission research was organized for the war effort, Seaborg moved in April 1942 from Berkeley to a new lab at the University of

Chicago known as the Metallurgical Laboratory (Met Lab) that was headed by physicist Arthur Compton. Seaborg's job there was to develop a method for separating plutonium from uranium in the reactor products. Physicist Enrico Fermi, who had worked on fission at Columbia University, was by this time also working at the Met Lab. On December 2, 1942, Fermi's team achieved the first self-sustaining nuclear reaction in a graphite and uranium pile called Chicago Pile-1 (CP-1). The pile, so named because uranium and uranium oxide lumps were stacked in a cubic lattice embedded in graphite, was built in a double's squash court located under the stands of the University of Chicago's Stagg Field. Plutonium co-discoverer Kennedy had left Berkeley to join the Project, moving in March 1943 to the Los Alamos site, where he eventually oversaw the Los Alamos Chemistry and Metallurgy Division (CM). Segrè joined him there and became head of the Site Y Radioactivity Group within its Experimental Physics Division.

Theoretical information from Fermi's CP-1 was the prototype for the experimental plutonium production reactor (X-10) at Clinton Engineer Works. The irradiated uranium was sent from X-10 to Clinton's chemical separation plant (Plant T) where plutonium-239 was separated from the irradiated uranium, using precipitation methods developed by Seaborg and chemists at the Met Lab. Information from the CP-1 and X-10 reactors and Clinton's chemical separation plant informed design of the industrial-scale B reactor and chemical separation plants at the Hanford site in eastern Washington, which furnished the mass amounts of plutonium "product" needed for the bomb.

POLONIUM

The uranium bomb's gun-type trigger assembly was a slow detonation system that could not be used in the plutonium bomb. The plutonium-240 isotope produced by irradiated uranium is highly prone to spontaneous emission of neutrons and would detonate before the bomb had triggered, causing a meltdown of the chain reaction. The plutonium bomb required, instead, a lightning-quick detonation system. Scientists turned to the implosion method, which used polonium to initiate a fission chain reaction. Polonium occurs naturally in lead residues or can be isolated mechanically from irradiated bismuth. It was first artificially produced by deuteron bombardment in 1936. Focused research on polonium for the bomb project was first undertaken at Berkeley and then on a larger scale in Dayton.

POLONIUM (INITIATORS) AS A TRIGGER

Polonium was used as the initiator, or "urchin," to trigger both the Fat Man bomb and the Little Boy bombs. The initiators supplied neutrons in a single burst to ensure that the chain reaction would start fast, and at exactly the right moment. The initiator designs for each bomb type were slightly different.

FAT MAN (PLUTONIUM—IMPLOSION METHOD)

In the implosion method, a fissile mass of uranium-235, plutonium-239, or a combination is surrounded by high explosives that compress the mass inward (implode) to the fissionable core, resulting in criticality. The Fat Man initiator was located in a pit at the center of the spherical bomb. It consisted of a hollow beryllium shell, with a solid beryllium pellet inside. Fifteen concentric latitudinal grooves were cut into the inner surface of the beryllium shell. The shell was formed in two halves that were coated with a layer of nickel and plated with a thin layer of gold. The gold and nickel layers protected the beryllium from alpha particles emitted by the 50 curies of polonium-210 (11 mg) that were deposited on the grooves inside the shell and on the central sphere.

As described by nuclear weapons historian Carey Sublette, hydrodynamic forces acting on the grooved shell mixed the beryllium and polonium, allowing the alpha particles from the polonium to impinge on the beryllium atoms. Reacting to alpha particle bombardment, the beryllium atoms emitted neutrons at a rate of about one neutron every 5 to 10 nanoseconds. These neutrons triggered the chain reaction in the compressed supercritical plutonium. Placing the polonium layer between two large masses of beryllium ensured contact of the metals, even if the shock wave turbulence performed poorly.[8] The urchin was activated by the arrival of the implosion shock wave formed by explosive lenses at the center of the core. When the shock wave from the implosion of the plutonium core arrived, it crushed the initiator.

LITTLE BOY (URANIUM—GUN-ASSEMBLY METHOD)

The uranium bomb relied on the gun-assembly design, which sent a subcritical uranium-235 projectile into a subcritical uranium-235 core. When the two met, the mass became supercritical. This initiator was simpler in design than that used in the plutonium bomb and contained less polonium. It was activated by the impact of the uranium projectile. The bomb would have exploded, even without the initiator.[9]

FURTHER INFORMATION

Technical information on the polonium work can be found in sources including the 36-volume government document *Manhattan District History*, in particular "Book VIII, Los Alamos Project (Y)—Volume 3, Auxiliary Activities, Chapter 4, Dayton Project"; Harvey V. Moyer, ed., "Polonium" (Oak Ridge: United States Atomic Energy Commission, Technical Information Service Extension, 1956)[10]; John Coster-Mullen's *Atom Bombs: The Top Secret Inside Story of Little Boy and Fat Man* (self-published, 2002); and Carey Sublette's website, "The Nuclear Weapon Archive" (nuclearweaponarchive.org/).

Historical information is presented in *Critical Assembly: A Technical History of Los Alamos During the Oppenheimer Years, 1943–45* by Lillian Hoddeson, Paul W. Henrikson, Roger A. Meade, and Catherine Westfall (Cambridge: Cambridge University Press, 1993). The book offers a thorough look at the implosion and polonium programs.

Bruce Cameron Reed's *The History and Science of the Manhattan Project* (Heidelberg: Springer, 2014) presents a general overview of Project physics and chemistry. An overall look at the chemistry work leading up to and through the war is described in William Albert Noyes Jr.'s "Offensive Chemical Warfare and Related Problems," found in *Science in World War II: Office of Scientific Research and Development, Chemistry: A History of the Science of Chemistry Components of the National Defense Research Committee, 1940–1946* (Boston: Little Brown & Co, 1948).

PROJECT-RELATED TRAVEL

NAME	FROM	TO	DATE
Otto Frisch (physicist)	Los Alamos	Dayton	March 30–April 3, 1944
Cyril Smith (metallurgist, associate director of Los Alamos Chemistry and Metallurgy)	Los Alamos	Dayton	June 30–July 4, 1943
		Chicago	Sept. 18–23
			Oct. 16–25
			Nov. 14–18
			Dec. 18–23
			Jan. 12–18, 1944
			Feb. 13–18
			May 14–19
			July 16–20
Rene Prestwood (chemist)	Los Alamos	Dayton	Oct. 10–17, 1943
Emilio Segrè (Los Alamos physics group leader)	Los Alamos	Chicago/ Dayton	Oct. 18–Nov. 5, 1943
Joseph Kennedy (chemist and head of Los Alamos chemistry and metallurgy)	Los Alamos	Berkeley	June 11–16, 1943
		Chicago	Oct. 17–21
			Nov. 14–19
			Dec. 19–22
			Jan. 12–15, 1944
			Feb. 13–18
			April 15–19
		Chicago/ Dayton	June 15–22

NAME	FROM	TO	DATE
		Chicago	July 16–20
		St. Louis	July 31–Aug. 3
		Chicago	Jan. 14–19, 1945 April 1–4 and April 26–28, 1945
J. Robert Oppenheimer (technical director of the Manhattan Project)	Los Alamos	Berkeley	June 11–16, 1943, consult with Thomas/Latimer
		Chicago/ Knoxville	Dec. 2–9
		Chicago	March 29–April 2, 1944
		Chicago/DC	May 15–22
		Chicago	July 15–19
		Chicago	July 23–25 with Groves Oct. 12–15
		Dayton/DC	May 28, 1945
Owen Chamberlain (physicist)	Los Alamos	Dayton	Dec. 27, 1943–Jan. 10, 1944
Richard W. Dodson (physicist, radiochemistry group leader)	Los Alamos	Dayton	July 9–14, 1944 April 9–19, 1945
George Kistiakowsky (physical chemist, explo- sive lenses/implosion)	Los Alamos	Dayton	Jan. 22–29 (approx.) "NDRC contract"
C. A. Thomas	Dayton	Los Alamos	May 31–June 4, 1943 Aug. 20–21 Sept. 23–25 Oct. 28–30 Dec. 16–18 Jan. 27–29, 1944 April 11–15 May 25–27 July 14–15 Jan. 29–31, 1945 July 17 Aug. 11–13 panel meeting
Carroll H. Hochwalt (Dayton Project assistant director)	Dayton	Los Alamos	July 20–25, 1943 Aug. 20–21

NAME	FROM	TO	DATE
Nicholas N. T. Samaras (Dayton Project assistant)	Dayton	Los Alamos	Sept. 6–8, 1943 Oct. 12–16 Nov. 15–21 Dec. 16–18 Feb. 22–26, 1944
James H. Lum (Dayton Project laboratory director)	Dayton	Los Alamos	April 11–15, 1944 July 2–5, 1945 Aug. 23–27

Source: Oppenheimer to Groves, "Arrivals and Departure of Visitors/Departures and Arrivals of Permanent Residents." May 24, 1943, and monthly thereafter.

DAYTON PROJECT PERSONNEL AND SELECT BIOGRAPHIES

A REVIEW OF Dayton Project personnel records from 1943 to 1946 shows staffing swelling to accommodate the increased requirements of the polonium work. There were 30 or so hires in 1943 as the Project began, compared to a marked increase of 116 or so in 1944 when the polonium work accelerated, and continued growth into 1945.

Hiring and termination dates of the employees were frequently close together—sometimes as short a stay on the job as one month—indicating that many people came and went from the project. These were support personnel, most likely not scientists, as research staff stayed with the Dayton Project until January or February of 1946 when wartime activities wrapped up. Similar short-term employment patterns were repeated across the country and at other Manhattan Project sites. During the war, many jobs on the home front were open, and people job-hopped as they sought positions that offered the best pay or working conditions. Some men were called up for duty. Others disliked the work environment.

Unless otherwise noted, basic biographical information was found in *MDH*, Book VIII, Los Alamos Project (Y)—Volume 3, Auxiliary Activities, Chapter 4, Dayton Project.

DAYTON PROJECT

End of 1944: 101 on payroll

End of 1945: 201 on payroll

Dec. 1946: 334 on payroll (37 administration; 49 research; 68 operations; 12 maintenance; 43 protection; 113 services; 12 craftsmen)

ADMINISTRATORS

Project director: Charles Allen Thomas (1900–1982): See Thomas vita, Appendix IV.

Assistant project director: Carroll A. Hochwalt (1899–1987), BA 1920, chemistry, University of Dayton; PhD 1935, chemistry, University of Dayton. Associate director of Monsanto Central Research Department. He was a research chemist at General Motors from 1920 to 1924, then production manager of the Ethyl Gasoline Corporation, 1924–25. Co-founder of Thomas & Hochwalt Laboratories in 1926, which became part of Monsanto in 1936. He was named director of Monsanto Central Research Laboratories in April 1945 and director of the Atomic Energy Commission's Mound facility. He was named vice president of Monsanto in 1947.

Laboratory director: James Henry Lum (1903–1996), BS 1925, chemical engineering, Pennsylvania State University; PhD 1932, Yale University. Prior to his work with the Dayton Project, Lum was a research chemist at Thomas & Hochwalt Laboratories and a Monsanto group leader. Following the war, he was named Monsanto assistant vice president in 1945, executive director of general administration and operations for Clinton Laboratories in 1946, managing director of Monsanto Chemicals Ltd. in Australia, and director of research and development of Monsanto's organic division. He retired from Monsanto in 1963.[1]

Assistant laboratory director: Willis Conard Fernelius (1905–1986), BA 1925, MA 1926, and PhD 1928, Stanford University. He joined the chemistry faculty at Ohio State University in 1928 and was named professor there in 1940, and then served as professor of chemistry at Purdue University (1942–47). He was on leave from Purdue with the Dayton Project from August 23, 1943, to August 6, 1946. After the war, he became chemistry department head at Syracuse University (1947) and then at Pennsylvania State University (1949–1960). He was also on the faculty at the University of South Florida and was adjunct professor of chemistry at Kent State University (1977–86). He was associate director of research, Koppers Company Inc., Monroeville, Pennsylvania.[2]

RESEARCH STAFF (PARTIAL LISTING)

Gösta Carl Akerlöf (1897–1966), PhD, Yale University, chemist and inventor. He left the physical chemistry faculty at Yale in June 1941 to join Thomas & Hochwalt Laboratories. He remained with the Dayton Project until October 1945 when he joined the Mellon Institute in Pittsburgh, supported by Koppers Chemical Corp. He then worked with the Naval Powder Factory in Maryland and moved to Princeton University's Forrestal Research Laboratories, followed by work in his own research laboratory.

Douglas Anger (1923–2005), BS, geology and chemistry, Colgate. He joined the Dayton Project as an Army SED chemist. Following the war, he earned a PhD in psychology from Harvard, where he studied with B. F. Skinner. He taught at Western Michigan University and did research at Upjohn Co. From 1970 until his retirement in 1989, he was a professor of psychology at the University of Missouri–Columbia.[3]

Catherine (Brenneman) Heyd (1915–1979) worked in the Unit III counting room. She married chemist and electronics section chief Josef Heyd.

Joseph J. Burbage (1914–1995), BS 1935, education, Miami University; MS and PhD, chemistry, the Ohio State University. He taught high school science for four years, and was a NDRC research fellow at Ohio State. He joined the staff in Dayton in late August 1943, becoming production manager of Unit IV in October 1945, then director of Mound Laboratory. In 1955, he was named director of development for the development and engineering department of Monsanto's inorganic chemicals division.[4]

Lowell V. Coulter (1913–2009), BA 1935, Heidelberg College; MA 1937, Colorado College; PhD, University of California, Berkeley). A physical chemist, he arrived in Dayton in May 1944 from Boston University and developed a calorimeter for assaying polonium. He returned to the BU chemistry faculty in October 1946 and was chairman from 1961 to 1973. He retired in 1977.[5]

Mary Lou Curtis (1911–2003), BS 1932, mathematics and English, minor in physics, Miami University; MA 1938, mathematics, Miami University. Curtis taught secondary algebra at New Richmond High School in Ohio from 1933 to 1938 and joined the Dayton Project on December 13, 1943. She was a physicist in the electronic counting department, where she pioneered the measurement of nuclear particles in radioactive isotopes. From 1946 to 1974 she worked for Mound Laboratory, where she was a research physicist, producing fuel cells for the U.S. space program, among other responsibilities. She also developed a method for calibrating and implemented the use of LOGAC, a low geometry alpha counter, one of the most accurate counting devices, best suited for routine alpha assay.[6]

Sergio De Benedetti (1912–1994), PhD 1933, physics, University of Florence, Italy. De Benedetti fled Mussolini's Italy and received a fellowship to the Curie laboratories in Paris, where he worked until 1940. When the Nazis invaded Paris, he fled that city on a bicycle, arriving in the United States via Portugal with a visa sponsored by Enrico Fermi and Arthur Compton. Once in the United States, he worked at the Bartol Foundation of the Franklin Institute in Philadelphia and then taught Army Air Recruits at Kenyon College. He arrived in Dayton on February 7, 1944, and as senior research physicist developed electronic counters and calorimetry equipment. He transferred to Clinton Laboratories in January 1946 and then joined the faculty at the Carnegie Institute of Technology (now Carnegie Mellon University), where he pioneered the use of positrons as probes to study properties of materials. He retired in 1984.[7]

John F. Eichelberger (1917–2009), a physicist who worked at Unit III. Following the war, he became a research director at Mound Laboratory.

G. Robert Gunther-Mohr (1922–2014), BS 1944, physics, Yale University; PhD 1954, physics, Columbia University. He joined the staff in 1944 as one of the first seven Army SEDs on the Dayton Project. After the war, he worked for thirty years in the semiconductors area of IBM and as a manager of physics and engineering at Watson Research Lab in Yorktown, New York.[8]

Betty (Halley) Jones joined in May 1945 and worked on construction of electronics instruments at Unit III. After completing studies at Roosevelt High School in Dayton, she attended Parker Cooperative School, where she took courses in sheet metal work, airplane mechanics, and drafting. Before arriving at the Dayton Project, she worked at Patterson Field on the flight line, servicing airplanes. Subsequent jobs included quartz crystal grinding for radio frequency crystal work at Aircraft Accessories Inc. and inspection of prisms for periscopes for Univis Lens. On the side, she modeled for portraits, hair, and clothing ads. Following the war, she joined Mound Laboratories and was sent to Los Alamos in 1955 to learn about detonator production. Upon her return, she set up training labs and production of detonators. She returned to Los Alamos in 1961 to help in the bending and cutting of the MDF (mild detonating fuze). Upon her return to the Ohio laboratory, she set up the timer program. In 1973 she was sent to Pantex in Amarillo, Texas, to change a detonator on a live bomb. These were regularly tested, and the bomb had a suspicious detonator. She removed the suspected one and replaced it with a good one. Halley was employed at Mound Laboratory until 1979.[9]

Malcolm M. Haring (1895–1952), BA 1915, chemistry, Franklin & Marshall College; MA 1917, chemistry, Princeton; PhD 1923, chemistry, Columbia University. He joined as a senior research chemist in August 1944, taking leave as a professor of physical chemistry from the University of Maryland. He became a group leader in fundamental research. He was named assistant laboratory director in July 1945, laboratory director in June 1946, and director of Mound Laboratories. He worked for Monsanto until his death in 1952.[10]

Josef Wilhelm Heyd (1908–1978), a research organic chemist with an electronics hobby, became electronics section chief after joining in January 1944. He remained with Monsanto until December 1969. He married counting room technician Catherine Brenneman.

Samuel S. Jones (1923–2014), BS 1943, chemistry, Hampden-Sydney College; PhD 1950, Cornell University. He was a physical chemistry graduate student at Cornell when he was drafted in the Army and sent to Dayton as a member of the Army SED. Jones returned to Cornell after the war, and in 1950 earned a PhD in chemistry, physics, and mathematics. He worked at General Electric, where he designed nuclear power systems for submarines, land-based plants, and space vehicles. In 1963, he moved to Richland, Washington, and worked on mechanisms for gas-graphite reactions and radiation effects on the properties of nuclear graphites. From 1970 to 1981, he was a senior staff research scientist at a variety of corporations including Kaiser Aluminum & Chemical Corp. focused on ways to reduce the cost of producing aluminum and other metals for industrial purposes. From 1983 until his retirement in 1996, he served as an industrial carbon consultant.[11]

Wilfred Konneker (1922–2016), BS 1943, chemistry; and MS 1947, atomic physics, Ohio University; PhD 1950, nuclear physics, Washington University. After completing undergraduate studies in chemistry at Ohio

University, Konneker joined the work in Dayton as a chemist with the Army SED. After the war, he studied nuclear physics and became a pioneer in nuclear medicine. He founded several companies in St. Louis, among them the radiopharmaceutical firm Nuclear Consultants, which was sold to Mallinckrodt Chemical Works in the 1960s. He ran the company's diagnostics division He was also director of Ohio University's Innovation Center from 1983 to 1991, and was co-founder of Embryogen Inc. and Diagnostic Hybrids Inc.[12]

Henry G. Kuivila (1917–2004), BS 1942 and MA 1944, Ohio State University; PhD 1948, Harvard. A chemist, he joined the project on March 15, 1944. He left in January 1946 and joined the chemistry faculty at the University of New Hampshire. In 1964, he became the first chairman of the department of chemistry at the State University of New York at Albany. He retired in 1988.[13]

Edwin M. Larsen (1915–2001), BS 1937, University of Wisconsin; PhD 1942, Ohio State University. An inorganic chemist on leave from the University of Wisconsin, he joined on September 13, 1943, and worked with Fernelius on the development of processes for purifying polonium. He married Dayton Project staff member Kathryn Behm in 1946 and returned to teaching at the University of Wisconsin. He was part of the Wisconsin Fusion Technology Institute, conducting research related to nuclear fusion power plants. He retired in 1986.[14]

Fred John Leitz Jr. (1921–2014), BA 1940, Reed College; PhD 1943, chemistry, University of California, Berkeley. The only member of the Dayton Project staff previously trained in radiochemistry, having written a dissertation at Berkeley on "The Photodisintegration of Some Light Elements." He arrived on January 4, 1944, as an Army SED with a knowledge of the bismuth extraction process, and initiated neutron counting work. Following the war, he transferred to Clinton Laboratories and later worked as a nuclear physicist focused on the production of nuclear power.[15]

Louis Eugene Marchi (1916–1994), BS 1938, Northwestern; PhD 1942, Ohio State. He joined on December 20, 1943, on leave from the Indiana University chemistry faculty. He left Dayton on February 8, 1946, and joined the chemistry faculty at the University of Pittsburgh.

Don S. Martin, BS 1939, Purdue University; PhD 1944, chemistry, California Institute of Technology. An inorganic chemist whose graduate work focused on radiochemistry, Martin worked on the volatility of polonium. Following the war, he worked at Iowa State College's Institute for Atomic

Research, directed by Frank Spedding who had been involved with the Manhattan Project. He was also on the chemistry faculty at Iowa State College.[16]

Edward C. McCarthy, a chemical engineer from the U.S. Rubber Company, joined the Dayton Project on January 27, 1944, to serve as chief of the design and engineering section. After the war, he served as plant manager for Mound Laboratory and was then appointed manager of the Monsanto inorganic chemical division's silicon development project in St. Louis. In 1959, he was named plant manager of the company's ultra-pure silicon metal production facility in St. Charles County, Missouri.[17]

Ralph E. Meints (1906–1988), PhD 1932, chemistry, University of Illinois, Urbana–Champaign. A chemist who came from the Army SED in the spring of 1944, he had a hand in the design of Unit IV, and was its first production manager. He left in fall 1945 to work at Argonne Laboratory in Chicago as chief design coordinator. He also taught chemistry at the University of Chicago (Chicago Circle) and was a lecturer in physical sciences at Roosevelt University in Chicago.[18]

Mary (Baluk) Moshier (1905–1999), BA 1929, University of Arkansas. She worked in the Monsanto technical library from 1936 to 1941, and then as a patent chemist (1942–1945). After receiving a law degree from Northern Kentucky University in 1962, she worked as a patent attorney. She contributed with her husband, chemist Ross W. Moshier, to the Charles A. Thomas monograph *Anhydrous Aluminum Hydride in Organic Chemistry* (American Chemical Society, 1941).

Ross W. Moshier (1899–1984), BS 1924, chemistry, Hillsdale College; MS Central Michigan University; PhD 1935, University of Michigan. A Monsanto employee, he joined the Project in October 1943 and worked in the analytical field, on electrolytic studies, and in extraction. He was the first to identify silver as a troublesome impurity in the purification process. After the war, he was a chemist in the Aeronautical Research Laboratories at USAF Wright-Patterson Air Force Base.[19]

Morris Lowell Nielsen (1914–2006), BS 1934, Yankton College; PhD 1941, chemistry, University of Wisconsin. Nielsen taught high school science, mathematics, and band for two years in Turton, South Dakota, before enrolling in 1938 at the University of Wisconsin. He was employed by Monsanto in Anniston, Alabama and then in Dayton before joining the Dayton Project as a chemist on October 18, 1943. He served as production manager for Unit IV. Following the war, he earned a law degree at the age of 50 from the Salmon P. Chase College of Law at Northern Kentucky University (1958), and worked as a patent attorney for Monsanto before moving to the Upjohn Company in Kalamazoo, Michigan. He retired in 1979. Among other accomplishments, he recorded scientific and legal books for the blind.[20]

Carl Linden Rollinson (1909–1995), BS 1933, University of Michigan; PhD 1939, University of Illinois. A process chemist from DuPont in Cleveland, he arrived in Dayton in November 1943. He left in August 1946 to teach chemistry at the University of Maryland.

Nicholas N. T. Samaras, BS 1927, chemical engineering, Case School of Applied Science; PhD 1931 Yale University. He remained at Yale for three years as a research fellow, then joined Thomas & Hochwalt Laboratories as a research chemist in 1934. He became a group leader for Monsanto Central Research and led the company's work on styrene for synthetic rubber. He joined the Dayton Project in 1943 as assistant research director, then in 1944 was named director of Monsanto's plastics division research laboratory in Springfield, Massachusetts.[21]

Walter J. Sampson (1909–1957) joined from Monsanto in November 1943 and was a structural engineer who planned the remodeling of what became Unit III, and designed much of the laboratory furniture. He transferred to Monsanto's Nitro plant in September 1944.

Cameron B. Satterthwaite (1920–2011), BA 1942, chemistry, College of Wooster; PhD 1951, chemistry, University of Pittsburgh. He left graduate studies at Ohio State University to join the Dayton Project in electrochemistry in 1944 and remained on staff until 1947. He worked for DuPont from 1950 to 1953 and Westinghouse from 1953 to 1961. He then joined the physics faculty at the University of Illinois, Urbana–Champaign, and from 1979 until 1985 was chair of the physics department at Virginia Commonwealth University.[22]

John W. Schulte (1919–2000), a chemist, he arrived on December 20, 1943 and remained in Dayton until June 1946. Following the war, he worked at Los Alamos Scientific Laboratory.[23]

D. L. Scott arrived in December 1943 from the Scioto Ordnance Plant of the U.S. Rubber Company and worked with design and application of sound production procedures.

Louis B. Silverman joined in October 1943. He worked in health physics and was a radiological safety engineer. He left Dayton in November

1945. Following the war, he joined the UCLA School of Medicine Atomic Energy Project.

John J. Sopka (1919–2014), BA 1942, physics, Harvard; PhD 1950, mathematics, Harvard. Following his undergraduate studies at Harvard, he worked at MIT's Radiation Laboratory and then taught introductory physics at Princeton to officers in the Navy V-12 College Training Program and enlisted men in the Army Specialized Training Program. He interrupted his physics doctoral studies at Harvard to become the Dayton Project's first staff physicist, joining the staff on January 17, 1944. He returned to Harvard in September 1945 and obtained a PhD in mathematics in 1950. He taught at Johns Hopkins before joining IBM, and then became director of computing for the National Bureau of Standards Laboratory in Boulder, Colorado. He was a professor of mathematics, first at the University of Texas in Arlington, Texas, later at Boston College, and finally at Fort Lewis College in Durango, Colorado.[24]

Joseph J. Spicka joined the project as a buyer on November 22, 1943, serving next as purchasing agent, and then as business manager. He remained with Monsanto until 1971.

Robert A. Staniforth (1918–1997), BA 1939, chemistry, Western Reserve University; MS 1942, chemistry, Ohio State University; PhD 1943, chemistry, Ohio State University. An inorganic chemist, he authored the book *Transuvanium Elements and Nuclear Fission* in 1942 and joined Dayton's Unit III in January 1944 on leave from General Aniline & Film Corp. He remained with the Atomic Energy Commission's Mound facility after the war as a group leader for fundamental research and chief of the chemistry research section from June 1946 to June 1947, a section chief from June 1947 to September 1948, and director of the research and development division beginning in October 1948. He developed a purity assay using an adapted Kirk-Craig balance.[25]

Eleanor (Stibitz) Billmyer (1918–2016), BA 1940, Heidelberg College. She was with the Dayton Project from December 29, 1943, to February 10, 1945. After the war, she worked as a newspaper reporter, editor, state public relations officer, and Albany County (NY) democratic legislator.[26]

Joseph Louis Svirbely (1906–1994), PhD 1931, University of Pittsburgh. A chemist and early pioneer on vitamin C research, he was a biochemist with the National Institute of Health and a specialist in industrial toxicology. He arrived in Dayton in November 1946 to head the health division and plan the biological research program. Svirbely had worked with the Mayo Clinic in Rochester, Minnesota, and had been a research fellow in the Department of Science at the Margaret Morrison Carnegie College, Carnegie Institute of Technology.

Harry R. Weimer (1906–1970), BS Manchester College; PhD 1933, the Ohio State University. He was on the faculty at Bridgewater College in Virginia for five years, then returned to Manchester College (Indiana) to teach and had been transferred to teach Navy personnel at Ball State University in Muncie, Indiana, when Fernelius recruited him to Dayton. He joined the Dayton Project on January 3, 1944, and helped develop the lead dioxide process. He returned to teaching in June 1945 and remained on the Manchester faculty until his death in 1970.[27]

Donald L. Woernley (1913–2003), PhD 1943, physics, Yale University. He joined the project as head of the physics section in 1943. Following the war, he joined the physics faculty at the University of Buffalo. He was named chief biophysicist at the Roswell Park Cancer Institute in Buffalo in 1949.[28]

Bernard S. Wolf (1912–1977), BS and MD, New York University. He was a radiologist at Mount Sinai Hospital in New York City when he was sent as a captain in the Army SED to Clinton Engineer Works. He arrived at the Dayton Project as health director in April 1945 and was transferred by the MED to the Madison Square Area in spring 1946, where he served as medical director for the Atomic Energy Commission. He was one of the first Americans to go to Hiroshima to monitor the effects of radiation. Following his work with the AEC, he became distinguished service professor at New York's Mount Sinai School of Medicine and chairman of the radiology department. He was also director of Mount Sinai Hospital's radiology department and chief radiologist. He pioneered ways in which radiology could be used to diagnose diseases of human organs.[29]

Richard Yalman (1923–), BS 1940, MS and PhD chemistry, 1948, Harvard University. He joined the Dayton Project in 1943 as a chemist, but after two months was drafted. After basic training, he returned to Dayton in May 1944 as an Army SED. After the war, he completed his education, then worked for Mound Laboratories for two years, before joining the faculty at Antioch College. He remained there until retirement in 1982. Following that, he and his wife owned an antique store in Santa Fe, New Mexico.[30]

OPERATIONS AND SERVICE STAFF

Sheridan J. Best joined in August 1944 and worked in mechanical maintenance.

Rachel Buck joined the Dayton Project on January 24, 1944.

Katie (Williams) Conway was a telephone operator.

Cleveland O. Dice arrived in January 1944 as master carpenter and supervisor of crafts.

Howard DuFour (1915–2009) was a machinist for the Dayton Project, and founder of the DuFour Machine Shop at Wright State University.

Howard E. Frost arrived in September 1943. The earliest non-technical employee, he worked as captain of the guards, supervisor of special maintenance, and cafeteria supervisor. He remained with Monsanto until April 1955.

Paul M. Hamilton joined the Dayton Project on February 1, 1944.

Edward Kerner was with the Dayton Project from February 29 to September 1, 1944.

C. H. Pittenger was supervisor of the machine shop. He joined on December 6, 1943, from Wright Field.

Pearl P. Rhodes was hired in February 1944 as a driver and became supervisor of transportation.

Evelyn Sands arrived from Wright Field in November 1943 as secretary to laboratory director James Lum and head of secretarial services. She was responsible for the supervision of Central Files.

George Timpe joined accounting in Unit III from Monsanto's St. Louis headquarters on October 18, 1943.

CHARLES ALLEN THOMAS VITA
February 15, 1900–March 29, 1982

EDUCATION

BA 1920, Transylvania University, Lexington, Kentucky
MS 1924, chemistry, Massachusetts Institute of Technology

HONORARY DEGREES

D.Sc. 1933	Transylvania University
D.Sc. 1947	Washington University
D.Sc. 1952	Princeton University
D. Sc. 1952	Kenyon College
D. Sc. 1953	Ohio Wesleyan University
D. Sc. 1956	Brown University
D. Eng. 1957	Polytechnic Institute of Brooklyn
D. Sc. 1958	University of Alabama
LL. D. 1950	Hobart College
LL. D. 1960	Lehigh University
D. Sc. 1965	St. Louis University
D. Eng. 1965	University of Missouri at Rolla
D. Sc. 1967	Simpson College

Westminster College, Fulton, Missouri

WORK

1923–24	Research chemist, General Motors Fuel Research Laboratory, Dayton, Ohio
1924–25	Research chemist, Ethyl Gasoline Corp., Dayton, Ohio
1926	Co-founder, Thomas & Hochwalt Laboratories, Dayton, Ohio
1928–36	President, Thomas & Hochwalt Laboratories, Dayton, Ohio
1936–45	Director, Monsanto Central Research, Dayton, Ohio
1942–43	Deputy chief, Section 8, National Defense Research Committee
1942	Member, Monsanto Chemical Company board of directors
1943	Vice president, Monsanto Chemical Company
1945–47	Director, Clinton Laboratories, Oak Ridge, Tennessee
1945	Vice president, technical director, and member of Monsanto Executive Committee
1946	Consultant to the scientific panel for the U.S. representative to the U.N. Atomic Energy Commission
1947	Executive vice president and member, Monsanto Finance Committee
1948	Vice chairman, Monsanto Executive Committee
1949	Chairman, Monsanto Executive Committee
1950–52	Chairman, Scientific Manpower Advisory Committee, National Security Resources Board
1951	Member, Presidential Scientific Advisory Committee, Office of Defense Mobilization
1951	President of Monsanto
1953	Consultant to the National Security Council
1955	Member, Business Advisory Council, U.S. Department of Commerce
1960–65	Chairman of the Board, Monsanto
1964–67	Chairman of the Technical Committee
1965–68	Chairman of the Finance Committee
1970	Retired from Monsanto

AWARDS

U.S. Medal for Merit, 1946

Industrial Research Institute Medal for achievements in the administration of
industrial research, 1947

Gold Medal, American Institute of Chemists recognizing work in research
administration, 1948

Elected to National Academy of Sciences, 1948

Missouri Award for Distinguished Service in Engineering, 1952

Deeds-Kettering Memorial Award, Dayton Engineers Club, 1964

Golden Plate Award, 1965

Elected to American Academy of Arts and Sciences, 1967

Perkin Medal, Society of Chemical Industries, American Section, 1953

Priestley Medal, American Chemical Society, 1955

Order of Leopold (Belgium), 1962

Palladium Medal, Société de Chimie Industrielle, 1963

St. Louis Globe-Democrat Man of the Year, 1966

William Greenleaf Eliot Society Search Award, Washington University, St. Louis,
Missouri, 1977

PROFESSIONAL ORGANIZATIONS

Director-at-large, American Chemical Society, 1942

President, American Chemical Society, 1948

Fellow, American Association for the Advancement of Science

Chairman, Board of Directors, American Chemical Society, 1950–53

Founding member, National Academy of Engineering, 1964

BOARDS

Chairman, Board of Trustees of Washington University, St. Louis

Chairman, Board of Directors, Washington University Medical Center

Curator, Transylvania College

Lifetime member, the Corporation of the Massachusetts Institute of Technology

Vice chairman, St. Louis Research Council

Trustee, Carnegie Corporation of New York

Trustee, John and Olga Queeny Educational Foundation
Board of Governors, National Farm Chemurgic Council
President, St. Louis United Fund, 1963
Leader of a $47 million bond issue campaign for the newly established St. Louis
Junior College District, 1965
Central Institute for the Deaf
Chemstrand Corporation
Civic Center Redevelopment Corporation of St. Louis
First National Bank, St. Louis
Metropolitan Life Insurance Company
RAND Corporation
Southwestern Bell Telephone Company
St. Louis Council of the Boy Scouts of America
St. Louis Union Trust Company

APPENDIX V

MANHATTAN PROJECT SITES AND PARTNERS (PARTIAL)

THE NUMBER OF academic, business, and industry participants in the Manhattan Project is staggering. It was truly a national effort, albeit one kept closely under cover. Subsidiaries such as Union Mines Development Corporation and Kellex were created to hide the bomb work. A listing of about 1,000 entities that contributed in some way to the project can be found in *Manhattan District History, Book 1, Volume 1, General* in an index titled "Agencies, Industrial, Organizations, Universities, etc." Here is a sampling:

PRIMARY SITES	
Los Alamos, New Mexico	Central laboratory
Clinton Engineer Works (Tennessee)	Uranium and plutonium enrichment facility
Hanford Engineer Works (Richland, Washington)	Plutonium production

AUXILIARY SITES	
University of Chicago Metallurgical Laboratory (Met Lab)	Plutonium chemistry and metallurgy
Iowa State College	Uranium
University of California, Berkeley (Radiological Laboratory/ RadLab)	Plutonium chemistry and metallurgy
Dayton, Ohio	Monsanto Chemical Company Polonium processing

PARTNERS

Grand Junction, Colorado	Uranium mining and processing. About 2.6 million pounds of uranium oxide (14 percent of the total acquired by the MED) was produced from this site.
Naval Proving Ground, Dahlgren, Virginia	Ordnance research and development
Explosives Research Laboratory, Bruceton, Pennsylvania	Impact sensitivity testing
Muroc Army Air Field, California (now Edwards AFB)	Testing of bomb delivery
Nash Garage Building, New York City	Site of Columbia University's pilot plant to create barrier material for K-25. The pilot was constructed by Kellex.
Naval Gun Factory, Washington, DC	Test guns for the development of the gun assembly device
Philadelphia Naval Yard	Research on liquid thermal diffusion method of isotope separation
Wright Field, Dayton, Ohio	Modification of B-29 Superfortress bomber under code name "Silverplate"
Wendover Army Air Base, Utah	Bombardment training base. Home of the 509th Composite Group and 216th, which delivered the bombs. Location of Project W-47, which assembled and tested models of Little Boy and Fat Man.

UNIVERSITIES

Brown University	Developed and manufactured crucibles for reducing plutonium to metal without introducing light-element impurities
California Institute of Technology	"Camel Project" to study weapon assembly mechanisms and combat delivery and to research and engineer specialized components including detonators
Columbia University	Home of Pupin Physics Laboratories and nuclear physicists Isidor Rabi, Enrico Fermi, Leo Szilard, George Pegram, John Dunning, and Harold Urey
Cornell University	Home of physicists Hans Bethe, Robert Bacher, Richard Feynman, and Col. Kenneth Nichols
Harvard University	Home of James B. Conant, Manhattan Project science advisor. Contributed cyclotron to Los Alamos.

Massachusetts Institute of Technology	Refractory
The Ohio State University	Cryogenic laboratory; properties and manufacture of liquid deuterium
Princeton University	Home of Eugene Wigner, Henry DeWolf Smyth, John von Neumann
Purdue University	Cyclotron, and fluorine research for Clinton's Y-12 plant
University of Michigan	Radar fuses and ordnance research
University of Minnesota	Electromagnetic method research
University of Rochester	Health physics research under Stafford Warren
University of Wisconsin	Provided Van de Graaff accelerators and physicists to Los Alamos

INDUSTRY

A. O. Smith Company, Milwaukee, Wisconsin	Heat exchangers to remove the heat of compression and piping that could withstand hexafluoride
Allis-Chalmers, West Allis, Wisconsin	Compressors designed to handle uranium hexafluoride at K-25
Batelle Memorial Institute, Columbus, Ohio	Studied fabrication of uranium and was a later site of the first nuclear research center
Chapman Valve Company, Massachusetts	Valve manufacture for Clinton Y-12 and machined uranium rods into slugs for the Atomic Energy Commission
Chrysler Corporation, Detroit, Michigan	Manufactured the converters (the corrosion-resistant tanks enclosing the diffusion barriers) for Clinton's K-25
Crane Manufacturing Company	Specialized valves for K-25
E. I. DuPont de Nemours and Company, Wilmington, Delaware	Operated Clinton plutonium production and then built and operated Hanford
Fercleve Corp.	Subsidiary of the H. K. Ferguson Construction Company of Cleveland that operated S-50 Liquid Thermal Diffusion Plant
Firestone Tire & Rubber Company, Akron, Ohio	Lead dioxide process for polonium retrieval
Ford, Bacon & Davis, Inc.	Worked with J. A. Jones Construction to build K-25

Hercules Powder Company	Experimental detonators
H. K. Ferguson, Inc., Cleveland, Ohio	An industrial construction and engineering firm contracted to design and build Clinton's S-50 liquid thermal diffusion plant, which produced enriched uranium
Houdaille-Hershey Co. Oakes Products Plant, Decatur, Illinois	Manufactured diffusion barriers for K-25
International Nickel Co., Huntington, West Virginia	Supplied nickel powder for K-25
J. A. Jones Construction, Charlotte, North Carolina	Managed construction of K-25 at Clinton
Joslyn Manufacturing and Supply Company, Fort Wayne, Indiana	Manufactured uranium rods for the Met Lab
W. E. Pratt Manufacturing Co., Joliet, Illinois	A subsidiary of Joslyn Manufacturing and Supply Co., Pratt began machining uranium slugs for the first reactors built at the University of Chicago. In 1944, Pratt also machined uranium rods for the Metallurgical Laboratory.
Kellex Corporation, Jersey City, New Jersey	A subsidiary of M. W. Kellogg Corp. created during the Manhattan Project. Kell(Kellogg)-X (secret) designed processes and equipment, and the K-25 gaseous diffusion reactor at Clinton.
Linde Air, Tonawanda, New York	Operated ceramics plant to process uranium ore and nickel
Lukas-Harold Corporation, Indianapolis	Produced barometric switch
Mallinckrodt Chemical Co. St. Louis, Missouri	Early uranium processing and waste management for the Manhattan Project
Monsanto Chemical Co., Dayton, Ohio	Dayton Project (1943–1947) Clinton Laboratories (1945–1947) Atomic Energy Commission/Department of Energy Mound Laboratory (1948–1988)
Norden Laboratories Corp.	Bomb altimeter
Quality Hardware and Machine Co., Chicago, Illinois	Produced experimental uranium fuel slugs in the summer of 1944
Roane-Anderson Co.	Managed the city of Clinton
Shepard-Niles Crane & Hoist	Crane hoist used at Trinity
Skidmore, Owens and Merrill	Architectural firm that designed the community of Oak Ridge

Stone & Webster Engineering Co.	Built Clinton's Y-12 electromagnetic separation plant and the community of Oak Ridge
Tennessee Eastman	Managed Clinton's Y-12
Union Carbide & Carbon Corp.	Operated Clinton's K-25 gaseous diffusion plant
Union Mines Development Corp.	A subsidiary of Union Carbide & Carbon Corp. created during the Manhattan Project to purchase uranium ore
Westinghouse	Developed centrifuges for refining uranium. This method was never used.
Whitlock Manufacturing Company	K-25: Heat exchangers to remove the heat of compression
Wycoff Drawn Steel Co., Chicago, Illinois	Machined uranium slugs for the Metallurgical Laboratory in 1943

Sources: blog.nuclearsecrecy.com/2013/05/24/inside-k-25/ and www.atomicheritage.org/history/project-sites

NOTES

NOTES TO THE PREFACE

1. *Manhattan District History (MDH)*, Book I—General, Volume 8, Personnel, 3.7.
2. The actual number of people hired by the Manhattan Project totaled nearly 600,000, or about one percent of the total U.S. wartime workforce when turnover from job terminations and resignations are taken into account. See Alex Wellerstein, "How Many People Worked on the Manhattan Project?," *Restricted Data: The Nuclear Secrecy Blog*, November 1, 2013, blog.nuclearsecrecy.co3/11/01/many-people-worked-manhattan-project/ (Accessed April 26, 2016).
3. A partial listing of industry and academic partners can be found in Appendix V, and in its entirety in *MDH*, Book I—General, Volume 1, 58–70.
4. "Monthly Report of the Chemistry and Metallurgy Division," March 1, 1945, Los Alamos Laboratory.
5. "Job and Personnel Summary for the CM Division." (Feb. 1945, and estimates for April 1945).
6. Kathren, Gough, and Benefiel, *The Plutonium Story*, 489.
7. See Appendix I for a review of literature.
8. The National Defense Authorization Act of 2015 included provisions authorizing the Manhattan Project National Historic Park to be located at three sites: Oak Ridge, Tennessee; Hanford, Washington; and Los Alamos, New Mexico. The Act was passed into law on December 19, 2014. Dayton was dropped from consideration early into the decade-long process of establishing the park.
9. "Historical Report—Dayton Area."
10. Roach, "How Man Handles It—Only the Future Will Know, Dayton and the Manhattan Project."

NOTES TO CHAPTER 1

1. LRG to Queeny (August 15, 1945).
2. Nichols, Interviewed by Groueff (January 4, 1965).

3. Staley, "Dayton, Ohio: The Rise and Fall and Stagnation of a Former Industrial Juggernaut."

4. "Brief History of Dayton" and "Dayton as an Industrial City."

5. Dalton, *Home Sweet Home Front: Dayton During World War II.*

6. "When Dayton Went to War: Memories of the Homefront," PBS; Staley, "Dayton, Ohio: The Rise and Fall and Stagnation of a Former Industrial Juggernaut."

NOTES TO CHAPTER 2

1. Bird, *Charles Allen Thomas, 1900–1982.*

2. DuPont, Speech of introduction at the Society of Chemical Industry's Perkin Medal presentation to Charles Allen Thomas.

3. "Musical Clubs Do Their Best at Spring Concert," *The Tech* (April 21, 1923): 6.

4. Thomas, *The Preparation of Benzene Sulphonyl Chloride and Some of Its Derivatives.*

5. CAT to FCT (August 15, 1923).

6. The Talbott family included Harold and Katharine Talbott and their children: Harold Jr., Daisy, Lillian, Nelson, Elsie, Marianna, Eliza, Katharine, and Margaret.

7. CAT to FCT (May 1924).

8. Hochwalt, Interview notes by Frank Curtis (December 29, 1950).

9. Hochwalt, Remarks introducing the medalist of Industrial Research Institute, "Charles Allen Thomas: The Man and the Scientist."

10. Koehnen, "Now and Then: Runnymede."

NOTES TO CHAPTER 3

1. Thayer, *Katharine Houk Talbott,* 310.

2. "Process of and Charge for Producing Carbon Dioxide at Low Temperatures," U.S. Patent 1,777,338, filed November 16, 1925, issued October 7, 1930; and "Charge for Fire Extinguishers," U.S. Patent 1,777,339, filed November 12, 1926, issued October 7, 1930. See also U.S. Patents 1,895,530; 1,895,691; 1,895,692; 1,910,653; 1,973,734; 2,063,772.

3. Thomas and Hochwalt, "Effect of Alkali-Metal Compounds on Combustion."

4. *Chemical & Engineering News* (January 5, 1948): 39.

5. Charles Allen and Mrs. Thomas, Interview notes by Frank Curtis, Monsanto, n.d.

6. Hochwalt, Transcript of interview by Sturchio and Thackray (July 12, 1985); Midgley, Thomas, Hochwalt, "Manufacture of Rubber," U.S. Patent number 1,806,547. Serial number 242,009, filed December 22, 1927, issued May 19, 1931.

7. Wright, "Monsanto's New President."

8. Hochwalt, Transcript of interview by Sturchio and Thackray (July 12, 1985).

9. Moshier, "Thomas and Hochwalt Laboratories Research Division of Monsanto Chemical Company."

10. Hochwalt, Transcript of interview by Sturchio and Thackray (July 12, 1985).

11. Charles A. Thomas and William H. Carmody published their resin research findings in the *Journal of the American Chemical Society*: "Polymerization of Diolefins

with Olefins: I. Isoprene and Pentene-2" (June 1932) and in *Industrial and Engineering Chemistry*: "Synthetic Resins from Petroleum Hydrocarbons" (October 1932).

12. Monsanto print advertisement: "Santopoid, Santolene, Santolube" (1939).

13. *St. Louis Post-Dispatch* (February 22, 1948).

14. Mr. and Mrs. Setzer, Interview notes by Frank Curtis (October 1951).

15. Martin interviewed by author (November 2005).

16. Charles Allen Thomas and Mrs. Thomas, Interview notes by Frank Curtis (February 8, 1951).

17. Hochwalt, Transcript of interview by Sturchio and Thackray (July 12, 1985).

18. Forrestal, *Faith, Hope & $5,000: The Story of Monsanto*, 85.

19. Mr. and Mrs. Setzer, Interview notes by Frank Curtis (October 1951).

20. CAT to FCT (October 24, 1933).

21. CAT III to Timothy Good, National Park Service (July 23, 2004).

22. "Charles A. Thomas, For a Job Done Well, Pick a Busy Man," *Chemical & Engineering News* (January 23, 1950): 24.

23. CAT III to author (September 18, 2005).

24. Charles Allen Thomas III (1927–2009), the only son and oldest child of Charles Allen Thomas, received a PhD in chemistry from Harvard University. He was a professor of biochemistry at Johns Hopkins University and Harvard Medical School before going into independent research in La Jolla, California.

25. Gluesenkamp and Mowry, Interview notes by Frank Curtis (October 1951).

26. *Fortune XI* magazine (May, June, and July 1935).

27. Ibid.

28. Forrestal, *Faith, Hope & 5,000: The Story of Monsanto*, 86.

29. "Corporations: Ready for Revolution," *Time* magazine.

30. Hochwalt, Interview notes by Frank Curtis (December 29, 1950).

31. Moshier, "Thomas and Hochwalt Laboratories, Research Division of Monsanto Chemical Company."

32. Hochwalt, Transcript of interview by Sturchio and Thackray (July 12, 1985).

33. Forrestal, *Faith, Hope & $5,000: The Story of Monsanto*, 83.

34. Hochwalt, Transcript of interview by Sturchio and Thackray (July 12, 1985).

35. A U.S. Patent application filed October 14, 1943, for "Resinous Condensation Product of Phenol and Styrene Oxide" described new oil-soluble resins for use in coating compositions, the formation of molded rigid articles of great density, and the production of friction materials. (U.S. Patent 2,422,637, Serial No. 506,265.)

NOTES TO CHAPTER 4

1. "Einstein Letter."

2. Rhodes, *The Making of the Atomic Bomb*, 331.

3. U.S. Library of Congress, Science Reference Services, Science, Technology & Business Division, Technical Reports and Standards, "The Office of Scientific Research and Development (OSRD) Collection," www.loc.gov/rr/scitech/trs/trsosrd.html.

4. Ibid.

5. Reed, *History and Science of the Manhattan Project*, 130–31.

6. Ibid., 132.

7. "Development of the OSRD."

8. "U.S. Synthetic Rubber Program," American Chemical Society.

9. Ibid.

10. Forrestal, *Faith, Hope & $5,000*, 74.

11. "U.S. Synthetic Rubber Program," American Chemical Society. See also "Rubber Matters, Solving the World War II Rubber Problem."

12. U.S. Library of Congress, "The Synthetic Rubber Project."

13. U.S. Patent 2,857,258 (Serial No. 612,133) for jet propellant was filed by Thomas and Monsanto on August 22, 1945, and patented October 21, 1958. U.S. Patent 3,014,796 (Serial No. 612,134), a related patent for a "new and improved" solid composite propellant, was filed by Thomas and Monsanto with Franklin A. Long and William M. Hutchinson representing the Secretary of War, on August 22, 1945. It was patented December 26, 1961.

14. Hochwalt, Remarks introducing the medalist of the Industrial Research Institute, "Charles Allen Thomas: The Man and the Scientist." See also Toedtman, "War Activities of the Central Research Department."

15. CAT to Captain Benton Bell, Division B, NDRC (April 2, 1943).

16. Kistiakowsky to CAT (April 17, 1943).

17. Bush to CAT (November 26, 1942).

18. For more information on the British effort, see Graham Farmelo's *Churchill's Bomb: How the United States Overtook Britain in the Nuclear Arms Race* (New York: Basic Books, 2013).

19. Frisch was Lise Meitner's nephew. He and Peierls joined Manhattan Project scientists at Los Alamos as part of the British delegation to Site Y.

20. Hewlett and Anderson, *The New World*, 261.

21. "FDR Approves Building an Atomic Bomb, 70th Anniversary October 9, 1941," The National WWII Museum. See also Hewlett and Anderson, *The New World*, 45.

22. Rhodes, "The Making of the Atomic Bomb," 386.

NOTES TO CHAPTER 5

1. CAT III interviewed by author, La Jolla, California (September 18, 2005).

2. Toedtman, "War Activities of the Central Research Department."

3. The Manhattan Engineer District was so-named by General Groves, who decided to follow the custom of naming Corps of Engineers districts for the city in which they are located, Source: www.atomicheritage.org/location/manhattan-ny. See also Richard Rhodes, *The Making of the Atomic Bomb*, 426.

4. Groves was promoted to the rank of Brigadier General shortly after his appointment to the Manhattan Engineer District and in 1944 was named Major General.

5. MED headquarters was moved to Oak Ridge, Tennessee, in August 1943, with Groves keeping his office in Washington, DC.

6. Lieutenant Colonel Kenneth D. Nichols was named deputy district engineer under Colonel James Marshall and then promoted to District Engineer of the Manhattan Engineer District. He was based in Oak Ridge.

7. Hewlett and Anderson, *The New World*, 291. See also "Staten Island's Tainted Edge," Geologic City Report #3.

8. The Tennessee facility's remote location in the Appalachian Mountains led scientists at the Met Lab in Chicago to call it "Down Under"; DuPont workers referred to it as "Gopher Training School"; and Monsanto dubbed it "Dogpatch," as reported in "High Flux Years," *Oak Ridge National Laboratory Review*, Numbers Three and Four, 30.

9. General Leslie M. Groves, *Now It Can Be Told*, 16.

10. When a uranium-235 atom absorbs a neutron and fissions into two new atoms, it releases three new neutrons and some binding energy. Two neutrons do not continue the reaction because they are lost or absorbed by a uranium-238 atom. However, one neutron does collide with an atom of uranium-235, which then fissions and releases two neutrons and some binding energy. Both of those neutrons collide with uranium-235 atoms, each of which fission and release between one and three neutrons, and so on. Source: "Science Behind the Atom Bomb," Atomic Heritage Foundation.

11. Smyth, *Atomic Energy for Military Purposes*, 6.4.

12. Kathren, Gough, and Benefiel, eds. *The Plutonium Story*, 189.

13. www.childrenofthemanhattanproject.org. To ensure secrecy, some 2,100 patents related to the Manhattan Project were filed secretly with the U.S. Patent Office but were not approved until decades after the war. See Alex Wellerstein's articles on the topic on his blog, *Restricted Data, The Nuclear Secrecy*, including "Patenting the Bomb: Nuclear Weapons, Intellectual Property, and Technological Control," *Isis* 99 No. 1 (March 2008): 57–87, blog.nuclearsecrecy.com/articles/.

14. Frank Settle, email communication with author (March 1, 2016).

15. U.S. Department of Energy, Office of History and Heritage Resources, "Difficult Choices (1942)," The Manhattan Project, an Interactive History.

16. Smyth, *Atomic Energy for Military Purposes*, 8.34–8.43.

17. Kathren, Gough, and Benefiel, *The Plutonium Story*.

18. AHC to Allison, Oppenheimer, Greenewalt, Seaborg (February 3, 1943).

19. By the summer of 1943, the divisions were: Chemistry and Metallurgy, with Joseph W. Kennedy, director, in charge of chemistry, and Cyril S. Smith, associate director, overseeing metallurgy; Theoretical, headed by Hans Bethe; Experimental, led by Robert F. Bacher; and Ordnance, guided by William S. Parsons. Kennedy and Smyth were officially appointed in January of 1944.

20. *MDH*: Los Alamos Project (Y), Book VIII, Volume 2 Technical, Chapter VIII, 8.13.

21. Hoddeson, Lillian et. al, *Critical Assembly: A Technical History of Los Alamos during the Oppenheimer Years, 1943–45*, 207; Hewlett and Anderson, *The New World*, 247.

22. "Manhattan Project History, Project Y, the Los Alamos Project," Los Alamos Scientific Laboratory, written 1946 and 1947, distributed December 1, 1961.

23. JRO to LRG (May 27, 1943).

24. Ibid.

25. RWD to JRO (April 30, 1945).

26. Hewlett and Anderson, *The New World*, 237.

27. Seaborg to CAT (October 18, 1968).

28. "Summary of the National Defense Research Council," 507.

29. Ibid.

30. Hoddeson et al., *Critical Assembly,* 121.

31. "Tribute to Dr. Thomas and Mr. Harrington," *St. Louis Post-Dispatch.*

32. "Lawrence and the Bomb," American Institute of Physics, History of Science Web Exhibits.

33. *St. Louis Post-Dispatch* (February 22, 1948).

34. Sanford, "Charles A. Thomas: A Voice for Atomic Control."

35. "Cheerful Evangelist in Atomics," *Science Illustrated,* 9.

36. JBC to CAT (July 28, 1942).

37. CAT to D. M. Sheehan, Monsanto (February 3, 1943).

38. U.S. Department of Energy, F. G. Gosling, "The Manhattan Project, Making the Atomic Bomb" (January 1999).

39. *MDH*: Dayton Project, "Historical Report: Dayton Project," 2.1.

40. JBC to CAT (July 27, 1944).

41. The work at the University of California, Berkeley, was undertaken by Wendell Latimer; at Iowa State College by Frank Spedding; in Chicago by Thorfin R. Hogness, Glenn Seaborg, and James Franck; and at Clinton by Warren Johnson.

42. Hochwalt, Remarks introducing the medalist of the Industrial Research Institute, "Charles Allen Thomas: The Man and the Scientist."

43. Kathren, Gough, and Benefiel, eds. *The Plutonium Story,* 289.

44. JBC to CAT (July 31, 1943).

45. Kathren, Gough, and Benefiel, *The Plutonium Story,* 293.

NOTES TO CHAPTER 6

1. Hammel, "The Taming of 49: Big Science in Little Time."

2. Kathren, Gough, and Benefiel, *The Plutonium Story,* 282.

3. Smith, "Plutonium Metal During 1943–45" from *The Metal Plutonium,* 29.

4. E. R. Jette to N. E. Bradbury (January 29, 1946).

5. JRO to LRG (August 31, 1944).

6. CAT to James Franck (October 1, 1943).

7. Hoddeson et al., *Critical Assembly,* 35; Conant to S-1 Executive Committee (November 20, 1942).

8. Hoddeson et al., *Critical Assembly,* 119.

9. Rhodes, *The Making of the Atomic Bomb,* 578.

10. Frank Settle, email communication to author (March 7, 2016).

11. *MDH*: Los Alamos Project (Y), Book VIII, Volume 2 Technical, Chapter IV, 4.21.

12. JRO to LRG (June 18, 1943).

13. Ibid.

14. Ibid.

15. One ton of uranium ore contains only about 100 micrograms (0.0001 grams) of polonium. Source: "It's Elemental," Jefferson Lab, education.jlab.org /itselemental/ele084.html; National Institute for Occupational Safety and Health, "Polonium from Irradiated Bismuth Slugs," in Evaluation Report Summary: SEC-00049 (May 2006), 5.1.2.

16. Joseph G. Hamilton to JRO (July 17, 1943).

17. Ibid.

18. National Institute for Occupational Safety and Health, "Evaluation Report Summary: SEC-00049, Monsanto Chemical Company."

19. A uranium hydride bomb was investigated during the early phases of the Manhattan Project but abandoned by early 1944. It used deuterium, an isotope of hydrogen, as a neutron moderator in a U-235-deuterium compound.

20. LRG to CAT (July 29, 1943).

NOTES TO CHAPTER 7

1. *MDH*: Dayton Project, 3.2.

2. Memorandum of Meeting (July 21, 1943).

3. A curie (Ci) is a unit of measurement of radioactive substance. One gram of polonium is the equivalent of about 6,000 curies, as reported by Oppenheimer to Groves on June 18, 1943.

4. Russian immigrant Boris Pregel brokered the deal through his Manhattan-based business, Canadian Radium and Uranium. He had supplied Fermi and Szilard with uranium for their fission experiments at Columbia. Groves was suspicious of Pregel's Russian background and later dropped him from contact with the bomb project.

5. Moyer, "Polonium," 116; JRO to LRG (June 18, 1943).

6. Ibid.

7. JRO to LRG (July 27, 1943).

8. CAT to LRG (August 28, 1943).

9. CAT to JRO (August 5, 1943).

10. JRO to CAT (August 5, 1943).

11. Ibid.

12. CAT to LRG and JBC (August 23, 1943).

13. Ibid.

14. Ibid.

15. "Sarasotan Recalls His Work on A-Bomb."

16. Osler, "Dayton's Wartime Secret Recalled."

17. *MDH*: Los Alamos Project (Y), Book VIII, Volume 2 Technical, Graph Number 1, "Age Distribution of Civilian Personnel, May 1945."

18. CAT to Herman B Wells (September 9, 1943).

19. CAT to JBC (September 21, 1943).

20. JBC to Wells (September 24, 1943).

21. Staniforth, MS Thesis, *Nuclear Fission and the Transuranium Elements*.

22. Staniforth quoted in a letter from E. C. Williams to Conant (December 6, 1943).

23. Sopka and Sopka, "The Bonebrake Theological Seminary: Top-Secret Manhattan Project Site."

24. See also Smith, "Plutonium Metal During 1943–45."

25. Ibid.

26. Shook and Williams, "Building the Bomb in Oakwood."

27. *MDH*: Dayton Project, 3.2.

28. Ibid.

29. Moyer, "Polonium," 215.

30. Hoddeson et al., *Critical Assembly,* 120.

31. Prestwood was in Dayton October 10–17, 1943. Source: Oppenheimer to Groves (May 24, 1943).

32. CAT to LRG and JBC (October 5, 1943).

33. In 1943, Thomas visited Los Alamos May 31–June 4, September 23–25, August 20–21, October 28–30, and December 16–18. In 1944, he was there January 27–29, April 11–15, and May 25–27, 1944. Reported by Oppenheimer to Groves (May 24, 1943).

34. Seaborg, quoted in "Charles Allen Thomas, 1900–1982" by Bird.

35. CAT to JRO (August 9, 1943).

36. *MDH*: Book VIII, Volume 2, Technical (Site Y), Chapter III, 3.13.

37. Norris, *Racing for the Bomb.*

38. Mr. and Mrs. Setzer, Interview notes by Frank Curtis (October 1951).

39. Personnel travel related to plutonium and polonium research is documented in Appendix II.

40. CAT III to author (2005).

41. "Monsanto Chemical Co. Will Produce Robot Launching Propellant."

42. *Dayton Journal* (October 21, 1944).

43. Shook and Williams, "Building the Bomb in Oakwood."

44. *MDH*: Dayton Project, 4.6.

45. Sopka and Sopka, "The Bonebrake Theological Seminary: Top-Secret Manhattan Project Site."

46. Ibid., 342.

47. Elisabeth Sopka, phone conversation with author (May 17, 2016).

48. The Manhattan Project included soldiers with technical training and scientific education who were drafted into the Army's Special Engineer Detachment. By 1945, there were 3,000 SEDs across MP sites.

49. Jones email to author (May 27, 2016); *The Mound Builder* (January 1953).

50. Weimer, "Harry and the Nuclear Threshold."

51. Weimer, Phone interview by author (March 23, 2016).

52. Weimer, "Harry and the Nuclear Threshold."

53. Bonnet, "Biography of Sergio De Benedetti."

54. Vera Bonnet (daughter of Sergio De Benedetti), Phone interview/email communication by author (December 12, 2015).

55. Emma De Benedetti, Phone interview by author (November 3, 2015).

56. Osler, "Dayton's Wartime Secret Recalled."

57. DuFour, Interview by James A. Kohler (May 17, 2006).

58. Ibid.

59. Shook and Williams, "Building the Bomb in Oakwood."

60. Seaborg to CAT (October 18, 1968).

61. Dayton Unit II, located near Ohio State Road 741, was used for work with the Army Air Command involving jet propellant for heavy aircraft. It was not related to the bomb.

62. Curtis interview, Atomic Heritage Foundation.

63. Kathren, Gough, and Benefiel, *The Plutonium Story*, 325.

64. "Conference on Polonium" at Ryerson Laboratory, University of Chicago (October 20, 1943); LRG to CAT (September 29, 1943).

65. CAT to AHC (October 1, 1943).

66. JRO to CAT (September 30, 1943).

67. Kathren, Gough, and Benefiel, *The Plutonium Story*, 365.

68. 10 curies would be 2230 ug or 2.23 milligrams.

69. CAT to LRG (November 3, 1943).

70. Between November 1943 and May 1945, 73,774 pounds (dry weight) of lead dioxide were received and processed at Unit III. By 1947, when the process ended, a total of 37 tons of lead dioxide was received and processed in Dayton, resulting in about 40 curies of polonium-210.

71. JRO to CAT (November 16, 1943).

72. Weimer, "Harry and the Nuclear Threshold."

73. CAT to LRG/JBC (October 5, 1943).

74. JRO, "Minutes of a Meeting on Polonium, December 16, 1943."

75. Curtis interview, Atomic Heritage Foundation.

76. Ibid.

77. CAT to JRO/JBC, "Report No. 3" (November 6, 1943).

78. Hoddeson et al., *Critical Assembly*, 123.

79. JRO to LRG (December 18, 1943).

80. "Minutes of a Meeting on Polonium, December 16, 1943" (December 18, 1943).

81. CAT to JRO/JBC, "Report No. 4" (January 4, 1944).

82. Ibid.

83. Ibid.

84. Toedtman, "War Activities of the Central Research Department."

NOTES TO CHAPTER 8

1. *MDH*: Dayton Project, 5.2.

2. Moyer, "Polonium," 3.

3. CAT to LRG/JBC, "Report No. 5" (February 10, 1944).

4. National Institute for Occupational Safety and Health, "Evaluation Report Summary."

5. Ibid.

6. Yalman interview, Atomic Heritage Foundation.

7. Ibid.

8. *MDH*: Dayton Project, 5.5.

9. Yalman interview, Atomic Heritage Foundation.

10. *MDH*: Dayton Project, 5.5. See also Moyer, "Polonium," 126.

11. Kathren, Gough, and Benefiel, *The Plutonium Story*, 365.

12. JRO to LRG (June 18, 1943).

13. Hoddeson et al., *Critical Assembly*, 124.

14. CAT to LRG/JBC, "Report No. 5" (February 10, 1944).

15. U.S. vs. Talbott Realty Company, District Court of the United States in and for the Southern District of the State of Ohio, No. 319 Civil (March 10, 1944).

16. Ibid.

NOTES TO CHAPTER 9

1. *MDH*: Dayton Project, 4.6 and 4.7.

2. Osler, "Dayton's Wartime Secret Recalled."

3. Ibid.

4. *MDH*: Dayton Project, 4.3.

5. Shook and Williams, "Building the Bomb in Oakwood."

6. Dooley, "Dooley Observed" blog.

7. Rosensweet, "Runnymede Playhouse Sold to U.S.; Will be Torn Down."

8. JBC to CAT (February 10, 1944).

9. CAT to LRG/JBC, "Report No. 6" (April 8, 1944).

10. Ibid.

11. CAT to LRG (April 6, 1944).

12. Nichols, Interview—Part I, Atomic Heritage Foundation.

13. Sopka and Sopka, "The Bonebrake Theological Seminary: Top-Secret Manhattan Project Site."

14. Shook and Williams. "Building the Bomb in Oakwood."

15. *MDH*: Dayton Project," 4.6.

16. Dudley, "Our Boys and the Bomb."

17. Roach, "How Man Handles It—Only the Future Will Know: Dayton and the Manhattan Project."

18. Franklin to Hochwalt (January 19, 1949).

19. Shook and Williams, "Building the Bomb in Oakwood."

20. CAT to LRG/JBC, June 13, 1944.

21. Thomas, "Science and Industry, A Powerful Team."

22. Ibid.

23. Thomas, "Radioisotopes—New Tools for Science."

24. Ibid.

25. Thomas, "Science and Industry, a Powerful Team."

26. Curtis interview, Atomic Heritage Foundation.

27. Ibid.

28. "Sarasotan Recalls His Work on A-Bomb."

29. 2.5 curies is a small amount, but this quantity of Polonium emits thousands of neutrons per second, enough for the triggering device.

30. CAT to JRO (February 24, 1944).

31. JRO to CAT (February 24, 1944).

32. JRO to CAT (March 14, 1944).

33. Kennedy, "Monthly Progress Report of the Chemistry and Metallurgy Division" (April 1, 1944).

34. CAT to LRG/JBC (April 8, 1944).

35. Moyer, "Polonium," 6.

36. RWD to JRO (August 23, 1944).

37. Smyth, *Atomic Energy for Military Purposes*, 7.3.

38. CAT to LRG/JBC (June 13, 1944).

39. The Military Policy Committee replaced the S-1 Executive Committee in 1943, though Bush, Conant, and other OSRD leaders remained on the new committee; Hewlett and Anderson, *The New World*, 245.

40. Alpha, beta, and gamma particles are emitted by radioactive material during decay. Alpha particles are fast-moving and high-energy helium atoms that, due to their large mass can, be stopped by paper. Beta particles are fast-moving electrons of moderate energy levels that can penetrate plastic or light metals. Gamma particles are high-energy photons that can penetrate lead.

41. Hoddeson et al., *Critical Assembly*, 228. See also Hammel, "The Taming of '49,'" *Los Alamos Science*.

42. When alpha particles strike nuclei of light element impurities, neutrons are generated in the resulting reaction. If the plutonium contains light element impurities generated from chemical processing, the weapon can pre-detonate due to the alpha (η) or (α,η) reactions arising from the impurities emitted by the Pu-240.

43. The Chemistry and Metallurgy Group Advisory Board was chaired by Arthur Compton and included Conant, Jeffries, Thomas, and Fermi.

44. Emilio Segrè's personnel report (June 13, 1945).

45. Kathren, Benefiel, and Gough, *The Plutonium Story*, 487.

46. Ibid.

47. JRO to LRG (July 18, 1944).

48. The neutrons would result from the reactions of alpha particles with impurities in the plutonium.

49. Kathren, Gough, and Benefiel, *The Plutonium Story*, 487.

50. Hewlett and Anderson, *The New World*, 251.

51. Fermi moved from the Met Lab to Site Y in September 1944 to direct the new F Division. He also became associate director of the laboratory and was placed in charge of theoretical and nuclear physics research.

52. Op. Cit.

53. CAT to LRG (July 21, 1944).

54. Kathren, Gough, and Benefiel, *The Plutonium Story*, 488.

55. CAT to LRG (July 21, 1944).

56. Kathren, Gough, and Benefiel, *The Plutonium Story*, 499; The editorial board consisted of John Chipman, Ermon D. Eastman, Thorfin R. Hogness, Joseph W. Kennedy, Glenn T. Seaborg, Cyril S. Smith, and Frank H. Spedding.

57. Sopka and Sopka, "The Bonebrake Theological Seminary: Top-Secret Manhattan Project Site."

58. *MDH*: Los Alamos Project (Y), Book VIII, Volume 2 Technical, Chapter IV, 4.53.

59. Reed, *The History and Science of the Manhattan Project*, 287.

60. Hewlett and Anderson, *The New World*, 253.

61. JBC to CAT (July 27, 1944).

62. JBC to JRO (July 27, 1944).

63. CAT to JBC (August 2, 1944). See also Hoddeson et al., *Critical Assembly*, 133.

64. For the initial work, some 2,000 pounds of bismuth was purchased from the American Smelting and Refining Co.

65. *MDH*: Dayton Project, 5.6. The Madison Square Area Engineers Office in New York City handled procurement of raw materials for the Project.

66. Ibid., 5.5.

67. A series of laboratory reports issued in 1948 by Monsanto Chemical Company contain the title "Gamma Scale Chemistry Progress Report." See also M. Economides, E. Estabrook, and E. F. Joy, "Gamma Scale Chemistry Progress Report," Monsanto Chemical Company—Unit III.

68. *MDH*: Dayton Project, 6.6.

69. Ibid.

70. Moyer, "Polonium," 150–52.

71. Ibid.

72. Former Mound employee Paul Lamberger, quoted in Ohio Environmental Agency Report (1998).

73. Moyer, "Polonium," 153.

74. Shook and Williams, "Building the Bomb in Oakwood."

75. T. S. Chapman to JRO (May 21, 1945).

76. Moyer, "Polonium," 317.

77. *MDH*: Dayton Project, 5.5

NOTES TO CHAPTER 10

1. Shook and Williams, "Building the Bomb in Oakwood."

2. Moyer, "Polonium," 346.

3. CAT to LRG/JBC, "Report No. 5" (February 10, 1944).

4. Smyth, *Atomic Energy for Military Purposes*, 8.67.

5. Unpublished article by Robert S. Norris.

6. Ibid.

7. "Monsanto Chemical Company—Unit III, Progress Report, September 1–15, 1945."

8. "Monsanto Chemical Company—Unit III, Progress Report, 10/15–10/31/45."

9. Lota, *The GRU and the Atomic Bomb* (2002). See also Walsh, "George Koval: Atomic Spy Unmasked," 40–47; Broad, "A Spy's Path: Iowa to A-Bomb to Kremlin Honor," 1–2; and Pagano, "The Spy Who Stole the Urchin: George Koval's Infiltration of the Manhattan Project."

10. DeBrosse, "Russian Spy Lived in Dayton, Stole Secrets."

11. *MDH*: Book I, Volume 14, Intelligence & Security, Foreign Intelligence, Part I, 4.8.

12. Sopka and Sopka, "The Bonebrake Theological Seminary: Top-Secret Manhattan Project Site."

13. RWD, "Notes on Discussions about Polonium, Dayton Ohio—April 11 to 14, 1945."

14. Shook and Williams, "Building the Bomb in Oakwood."

15. Yalman interview, Atomic Heritage Foundation.

16. *MDH*: Los Alamos Project (Y), Book VIII, Volume 2 Technical, Chapter IX, 9.33.

17. "Monthly Progress Report of the Chemistry and Metallurgy Division" (June 1, 1945).

18. Mahfouz, Interviewed by Roach and Good (October 7, 2004).

19. Ibid.

20. Sopka and Sopka, "The Bonebrake Theological Seminary: Top-Secret Manhattan Project Site."

21. Ibid.

22. Curtis interview, Atomic Heritage Foundation.

23. National Institute for Occupational Safety and Health, "Evaluation Report Summary: SEC-00049, Monsanto Chemical Company."

24. Hewlett and Anderson, *The New World*, 378

25. Stafford Warren to CAT (June 23, 1945).

26. Rhodes, *The Making of the Atomic Bomb*, 580.

27. Kennedy, "Monthly Progress Report of the Chemistry and Metallurgy Division" (December 1, 1944).

28. "Monthly Progress Report of the Chemistry and Metallurgy Division" (March 1, 1945).

29. Albright, Interviewed by author (October 10, 2008).

NOTES TO CHAPTER 11

1. Kathren, Gough, and Benefiel, *The Plutonium Story*, 652.

2. CAT to Nichols (February 8, 1945).

3. "Monthly Progress Report of the Chemistry and Metallurgy Division" (June 1, 1945).

4. Information on the German bomb effort was obtained in 1943 by an espionage unit organized by General Groves and known as "Alsos." See Rhodes, *The Making of the Atomic Bomb*, 607.

5. JRO to Langer (May 21, 1945).

6. CAT to JRO (May 22, 1945).

7. JRO to CAT (June 6, 1945).

8. CAT to Nichols (February 8, 1945).

9. Nichols to CAT (May 16, 1945).

10. *MDH*: Los Alamos Project (Y), Book VIII, Volume 2 Technical, Chapter VII, 7.75

11. JRO to CAT (June 6, 1945).

12. Nichols to JRO (July 11, 1945).

13. Gilbert, "History of the Dayton Project," 14.

14. Kathren, Gough, and Benefiel, *The Plutonium Story*, 608.

15. *MDH*: Book I—General, Volume 14, Intelligence and Security, Foreign Intelligence, Part I, 5.9.

16. Gittler interview, Atomic Heritage Foundation.

17. Ibid.

18. *MDH*: Book I—General, Volume 14, Intelligence and Security, Foreign Intelligence, Part I, 5.9.

19. Gittler interview, Atomic Heritage Foundation.

20. Ibid.

21. Kathren, Gough, and Benefiel, *The Plutonium Story*, 293.

NOTES TO CHAPTER 12

1. Bainbridge, "Trinity."

2. *MDH*, Los Alamos Project (Y), Book VIII, Volume 2 Technical, Chapter X, 10.7.

3. In 18 months of investigation, some 20,000 experimental castings of experimental quality were produced and a much larger number rejected. *MDH*, Los Alamos Project (Y), Book VIII, Volume 2 Technical, Chapter XVI, 16.27.

4. Los Alamos Scientific Laboratory, Public Relations staff, *Los Alamos, Beginning of An Era, 1943–45*, 34; JRO to "All group leaders concerned" (June 14, 1945).

5. JRO to "All group leaders concerned" (June 14, 1945); Rhodes, *The Making of the Atomic Bomb*, 655.

6. Kistiakowsky interview, Atomic Heritage Foundation.

7. Sublette, "Section 8.0, The First Nuclear Weapons."

8. Los Alamos Scientific Laboratory, Public Relations Office, *Los Alamos, Beginning of an Era, 1943–45*, 46.

9. AHC to JRO (July 13, 1945).

10. Laurence, "Drama of the Atomic Bomb Found Climax in July 16 Test."

11. CAT to FCT (August 29, 1945).

12. Smith, "Comments on Trinity Test Shot Trip."

13. Lamont, *Day of Trinity*, 230–39.

14. Ibid.

15. Smith, "Comments on Trinity Test Shot Trip."

16. Kathren, Gough, and Benefiel, *The Plutonium Story*, 728.

17. Thomas, "Crossroads." (June 13, 1946).

NOTES TO CHAPTER 13

1. Rhodes, *The Making of the Atomic Bomb*, 743.

2. "Text of Truman Secret Weapon Announcement," *St. Louis Post-Dispatch*.

3. *Dayton Daily News*, final edition.

4. "Truman Reveals Mighty Weapon Now Being Used," *Dayton Herald*.

5. "Sarasotan Recalls His Work on A-Bomb."

6. Curtis interview, Atomic Heritage Foundation.

7. Jones, Email communication with author (May 27, 2016).

8. "2 Dayton Firms Help Produce Atomic Bomb," *Dayton Herald*.

9. Nichols to CAT (August 6, 1945).

10. LRG to Queeny (August 15, 1945).

11. LRG to CAT (August 15, 1945).

12. JRO to CAT (September 8, 1945).

13. Smyth, *Atomic Energy for Military Purposes*, vii.

14. Kathren, Gough, and Benefiel, *The Plutonium Story*, 749.

15. Smyth, *Atomic Energy for Military Purposes*, 6.3.

16. "Key War Plant Leaders Get Awards," *Dayton Herald*.

17. "Praise Given Monsanto's Varied War-Aid Program," *Dayton Daily News*.

18. Weimer, Phone interview by author (March 23, 2016).

19. Weimer, "Harry and the Nuclear Threshold."

20. Jones, Email communication with author (May 27, 2016).

21. Queeny to Monsanto Shareholders (September 8, 1945).

22. LRG to CAT (November 1, 1945).

23. Some 329 Medal for Merit awards were presented. In addition to Charles Allen Thomas, scientists receiving Medal for Merit awards included Samuel Allison, Robert Bacher, Hans Bethe, Arthur Compton, John Dunning, Enrico Fermi, Lawrence Hafstad, Joseph Kennedy, George Kistiakowsky, Ernest Lawrence, J. Robert Oppenheimer, Cyril Stanley Smith, Robert Stone, Harold Urey, and Eugene Wigner.

24. Truman, "Citation to Accompany the Award of the Medal for Merit to Dr. Charles Allen Thomas."

25. See Appendix III for biographies of select Dayton Project personnel.

26. Gates interview, Atomic Heritage Foundation.

27. Sopka and Sopka, "The Bonebrake Theological Seminary: Top-Secret Manhattan Project Site."

28. De Benedetti, Phone interview by author (November 3, 2015).

NOTES TO CHAPTER 14

1. Smith, "Behind the Decision to Use the Atomic Bomb: Chicago 1944–1945." (Alice Kimball Smith was married to Los Alamos chemist Cyril Smith, who was head of the metallurgical division.)

2. Hewlett and Anderson, *The New World*, 325; Bird, "Charles Allen Thomas."

3. Kathren, Gough and Benefiel, *The Plutonium Story*, 631.

4. Monsanto assumed the work in Clinton under Contract No. Y35–058-eng-71.

5. CAT to LRG, September 6, 1945.

6. *MDH*: Book I, General, Volume 4, Auxiliary Activities, Chapter 2, "Foundation of the National Laboratories," 4.12.

7. Ibid., 2.3–2.5.

8. JRO to CAT, September 8, 1945.

9. "T Brochure."

10. "What Goes On Here."

11. "The Miami Valley Atomic Energy Show Collection"; Vincent, "Atomic Energy Exhibit Opens at Miamisburg."

12. "2 Dayton 'Explosions'—Vote Returns and Atoms."

13. Helbert, "Valley Atomic Energy Show Opens With Bang."

14. "Mound Ohio, Site."

15. "T-Building brochure."

16. U.S. vs. Talbott Realty Company.

17. LRG to CAT. November 1, 1945.

18. Hewlett and Anderson, *The New World*, 535.

19. Lang, A Reporter at Large, *New Yorker.* 17 August 1946. Reprinted in the *St. Louis Post-Dispatch*, 25 August 1946.

20. Sanford, "Charles A. Thomas: A Voice for Atomic Control."

21. "Your Future in the Atomic Age."

22. Hewlett and Anderson, *The New World*, 536.

23. Sanford, "Charles A. Thomas: A Voice for Atomic Control."

24. Hewlett and Anderson, *The New World*, 538.

25. The United Nations Atomic Energy Commission included six permanent members—the United States, Britain, France, the Soviet Union, China, and Canada—and six rotating members.

26. Hewlett and Anderson, *The New World*, 559.

27. Sanford, "Charles A. Thomas: A Voice for Atomic Control." For more on international control, see Hewlett and Anderson, *The New World*, 531–79.

28. U.S. Department of State Office of the Historian, "Milestones: 1945–1952, The Acheson-Lilienthal & Baruch Plans, 1946."

29. "Curb Atom or UN Will Fail, Speaker Warns," *St. Louis Post-Dispatch*. See also "Atomic Energy Man's Slave, Expert Says in Washington U. Commencement Talk," *St. Louis Post-Dispatch*.

30. Hewlett and Duncan, *Atomic Shield 1947/1952: Volume II of a History of the United States Atomic Energy Commission*.

31. Monsanto news release (April 24, 1951).

32. Hochwalt, Remarks introducing the medalist of the Industrial Research Institute, "Charles Allen Thomas: The Man and the Scientist."

33. CAT III to Bird, (August 9, 1990).

34. Thomas, "The Future is Nearly Here."

35. "Resolution of the Corporation of the Massachusetts Institute of Technology on the Death of Charles A. Thomas, Life Member Emeritus."

36. Stockbridge, "Still Going Strong at 80."

37. Thomas, "Priorities in Science."

NOTES TO APPENDIX I

1. *Manhattan District History*, Los Alamos Project (Y), Book VIII, Volume 2 Technical, Chapter I, 1.30.

2. The Ames Group produced more than 2 million pounds of uranium for the Manhattan Project; Hoddeson et al., *Critical Assembly*, 31.

3. Kathren, Gough, and Benefiel, *The Plutonium Story*, 260.

4. Segrè, personnel report (June 13, 1945).

5. Plutonium has five "common" isotopes. One of these is Pu-239. A second isotope, Pu-240, also figured in the Manhattan Project.

6. Kathren, Gough and Benefiel, *The Plutonium Story*, 216.

7. Ibid.

8. Sublette, "Section 8.0, The First Nuclear Weapons."

9. Ibid.

10. The 36-volume *Manhattan District History (MDH)*, commissioned by General Leslie Groves in 1944, was declassified and made available to the public by the U. S. Department of Energy in July 2014. It is available electronically at www.osti .gov/opennet/manhattan_district.jsp.

NOTES TO APPENDIX III

1. "Sarasotan Recalls His Work on A-Bomb."

2. "W.C. Fernelius to Tour Southwest in October."

3. Douglas Anger, 1923–2005, obituary.

4. "Joseph J. Burbage—Personal History."

5. Koch, "Faculty Obituaries: A Teacher, Recruiter, and Leader."

6. Curtis interview, Atomic Heritage Foundation.

7. "From Generation to Generation."

8. Gerard Robert Gunther-Mohr, Obituary.

9. Jones, Interviewed by author; "Beauteous Red Head Knows Electronics, Models Hair Styles, and Studies Nights."

10. "Background of Malcolm M. Haring"; "UMD History Connections: Malcolm and the Manhattan Project."

11. "Our Boys and the Bomb"; Samuel Stimpson Jones, Dunkum Funeral Home.

12. "Wilfred Konneker, Nuclear Medicine Pioneer and Philanthropist, Dies at 93"; "Ohio University Mourns the Death of Wilfred R. Konneker."

13. Henry Gabriel Kuivila, Obituary.

14. "Memorial Resolution of the Faculty of the University of Wisconsin–Madison on the Death of Professor Emeritus Edwin Merritt Larsen" (May 6, 2002). Accessed October 2, 2016, www.secfac.wisc.edu/senate/2002/0506/1642(mem_res).pdf

15. Leitz, Interview by Cameron E. Foster-Keddie; *News Tribune* (Tacoma) obituary (July 9, 2014).

16. *Institute for Atomic Research*.

17. U.S. Department of Health and Human Services, "Special Exposure Cohort Petition" (January 9, 2006).

18. "University of Illinois: Chemistry, 1941–1951, staff, developments, courses and curricula, publications, doctorate degrees."

19. "Six Join Faculty for First Year."

20. Morris L. Nielsen, Obituary

21. "Played Important Role in Releasing Atomic Energy."

22. Cameron B. Satterthwaite, Obituary.

23. John Walter Schulte, Obituary.

24. Sopka and Sopka, "The Bonebrake Theological Seminary: Top-Secret Manhattan Project Site."

25. "Robert Arthur Staniforth (personal history)."

26. Eleanor Billmyer, Obituary.

27. Weimer, "Harry and the Nuclear Threshold."

28. Ibid.

29. "Dr. Bernard S. Wolf, 65, Radiologist and Professor at Mount Sinai School."

30. Yalman interview, Atomic Heritage Foundation.

BIBLIOGRAPHY

BOOKS, PERIODICALS, WEBSITES

"2 Dayton Firms Help Produce Atomic Bomb." *Dayton Herald* (August 7, 1945).

"2 Dayton 'Explosions'—Vote Returns and Atom." *Dayton Daily News* (November 3, 1948).

"170-Acre Tract Bought as Site for New Plant." Newspaper unknown (n.d.). Mound Science and Energy Museum archive, Miamisburg, Ohio.

"1200 Pupils See Exhibit at Atom Show." *Dayton Journal* (April 30, 1948).

"100,000 Private Theater Auditorium Being Erected by Mrs. H. E. Talbott." Newspaper unknown (n.d.). Oakwood Historical Society, Dayton, Ohio.

"American Chemical Industries, Thomas & Hochwalt Laboratories, Inc.," *Industrial and Engineering Chemistry* 14 (May 10, 1936): 181. Monsanto Company Records, Washington University Libraries, Department of Special Collections.

"Architect's Drawing is First Released Showing Layout of Mound Hill Project." *Miamisburg News* (July 7, 1947).

"Atomic Authority Urges U.S. to Retain Control." *Dayton Daily News* (n.d.). Mound Science and Energy Museum archives, Miamisburg, Ohio.

"Atomic Bomb Makers Free at Last." *Dayton Journal* (August 8, 1945, late city edition).

"Atomic Bomb, Most Destructive Force in History, Hits Japan." *Dayton Daily News* (August 6, 1945, final edition).

"Atomic Energy for Power Plants Seen for Future." *Troy (Ohio) Daily News* (May 12, 1948).

"Atomic Energy Group to Vacate Playhouse Here." Newspaper unknown (n.d.). Mound Science and Energy Museum archives, Miamisburg, Ohio.

"Atomic Energy Man's Slave, Expert Says in Washington U. Commencement Talk." *St. Louis Post-Dispatch* (June 13, 1946).

"Atomic Energy Show Opens Tonight." *Dayton Herald* (November 3, 1948).

"Atom Smasher." *Time* magazine (August 13, 1945).

Babcock, Jim. "Dayton Lab Led the Way." *Dayton Daily News* (September 28, 1986).

"Background of Malcolm M. Haring." Miami Valley Energy Show Collection, Wright State University Libraries, Special Collections and Archives.

Baredoll, Robert. "July 16—12 Men Test Atomic Bomb." *Dayton Journal* (August 7, 1945, late city edition).

Barnard, Chester I., Dr. J. R. Oppenheimer, Dr. Charles A. Thomas, Harry A. Winne, David E. Lilienthal, Chairman. "A Report on the International Control of Atomic Energy." Washington, DC: Secretary of State's Committee on Atomic Energy, Publication 2498, March, 16, 1946.

"Beauteous Red Head Knows Electronics, Models Hair Styles, and Studies Nights." *The Mound Builder* 2, no. 4 (January 1953). Monsanto Chemical Company, Mound Laboratory.

Bernhard, Blythe. "Wilfred Konneker, Nuclear Medicine Pioneer and Philanthropist, Dies at 93." *St. Louis Post-Dispatch* (February 5, 2016). Accessed April 13, 2016, www .stltoday.com/news/local/education/wilfred-konneker-nuclear-medicine-pioneer -and-philanthropist-dies-at/article_4e9c3d92-b708-54e7-bf91-4eaf8037daae.html.

"Biographical Sketch of Charles Allen Thomas." Monsanto Company (May 1967). Charles Allen Thomas Papers, Washington University Libraries, Department of Special Collections.

Bird, Kai and Martin Sherwin, *American Prometheus, the Triumph and Tragedy of J. Robert Oppenheimer.* New York: Vintage Books, reprint 2006.

Bird, R. Byron. "Charles Allen Thomas: February 15, 1900–March 29, 1982." *Biographical Memoirs* 65 (1994). The National Academy of Sciences.

Bonnet, Vera De Benedetti. "Biography of Sergio De Benedetti." Sergio De Benedetti Manuscript. Book 3 (2015). digitalcommons.chapman.edu/debenedetti/3.

"Brief History of Dayton," City of Dayton, City Commission Office, www.cityofdayton .org/cco/Pages/BriefHistory.aspx. Accessed August 31, 2015.

Broad, William J. "A Spy's Path: Iowa to A-Bomb to Kremlin Honor." *The New York Times* (November 12, 2007).

"Buildings Used Here and Men Who Worked on Atomic Bomb." *Dayton Daily News* (n.d.): 23.

"Cameron B. Satterthwaite." *Physics Illinois*, In Memoriam. Accessed November 14, 2016, physics.illinois.edu/people/memorials/satterthwaite.

"Charles Allen Thomas," notes from *Chemistry and Engineering News* (January 5, 1948): 39. Charles Allen Thomas Papers, Washington University Libraries, Department of Special Collections.

"Charles Allen Thomas, Monsanto's New President." *Monsanto* magazine (June 1951): 4.

"Cheerful Evangelist in Atomics." *Science Illustrated* (March 1947).

"Corporations: Ready for Revolution." *Time* magazine (May 10, 1948).

Coster-Mullen, John. *Atom Bombs: The Top Secret Inside Story of Little Boy and Fat Man* (2002).

"Curb Atom or UN Will Fail, Speaker Warns." *St. Louis Post-Dispatch* (June 14, 1946).

"Current Projects—June 1945." Monsanto Chemical Company. Monsanto Company Records, Washington University Libraries, Department of Special Collections.

Dalton, Curt. *Home Sweet Home Front: Dayton During World War II*, n.p.: 2002.

"Dayton as an Industrial City." Dayton History Books Online, accessed August 31, 2015, www.daytonhistorybooks.com/page/page/1512274.htm.

Dayton Daily News, final edition, (August 6, 1945).

DeBrosse, Jim. "Russian Spy Lived in Dayton, Stole Secrets." *Dayton Daily News* (February 15, 2010).

Deep, Shannon. "From Generation to Generation," *Carnegie Mellon Today* (June 30, 2013).

"Developed Bomb in 3 States." *Dayton Daily News* (August 6, 1945, final edition).

Dooley, Joe. "Runnymede Playhouse." Joe Dooley's Dooley Observed blog (May 23, 2014). Accessed March 30, 2016, dooleyobserved.blogspot.com/2014/05/runnymede -playhouse.html.

"Douglas Anger, 1923–2005." *Columbia Daily Tribune* obituary (December 14, 2005). Accessed June 3, 2016, archive.columbiatribune.com/2005/Dec/20051214Obit001 .asp.

"Dr. Bernard S. Wolf, 65, Radiologist and Professor at Mount Sinai School." *New York Times* (September 17, 1977). Accessed June 1, 2016, www.nytimes.com/1977/09 /17/archives/dr-bernard-s-wolf-65-radiologist-and-professor-at-mount-sinai.html.

"Dramatic Show Unfolds Secrets of Mighty Atom." Newspaper unknown (n.d.). Mound Science and Energy Museum archives, Miamisburg, Ohio.

Dudley, John. "Our Boys and the Bomb." *The Record* (July 2010). Hampden-Sydney College. Accessed April 30, 2016, www.hsc.edu/TheRecord.

DuPont, Henry B. Speech of introduction at the Society of Chemical Industry's Perkin Medal presentation to Charles Allen Thomas. Waldorf-Astoria Hotel, New York City (January 16, 1953). Charles Allen Thomas Papers, Washington University Libraries, Department of Special Collections.

"Einstein Letter." Franklin D. Roosevelt Presidential Library and Museum, www .fdrlibrary.marist.edu/archives/pdfs/docsworldwar.pdf.

Eleanor Billmyer obituary. *Syracuse Post-Standard* (April 1–3, 2016). Accessed November 14, 2016, obits.syracuse.com/obituaries/syracuse/obituary.aspx?pid=179486272.

Farmelo, Graham. *Churchill's Bomb: How the United States Overtook Britain in the Nuclear Arms Race.* New York: Basic Books, 2013.

"FDR Approves Building an Atomic Bomb, 70th Anniversary October 9, 1941," The National WWII Museum. Accessed May 16, 2016, www.nww2m.com/2011/10 /fdr-approves-building-an-atomic-bomb-70th-anniversary-october-9-1941/.

Findlay, John and Bruce Hevly. *Atomic Frontier Days, Hanford and the American West.* Seattle: Center for the Study of the Pacific Northwest in association with the University of Washington Press, 2011.

"Flash Seen for 150 Miles When Atom Bomb Explodes." *Dayton Herald* (August 9, 1945, final edition).

Forrestal, Dan J. *Faith, Hope & $5,000: The Story of Monsanto.* New York: Simon & Schuster, 1977.

Fred John Leitz Jr. obituary. *News Tribune* (Tacoma, Washington) (July 9, 2014).

Gerard Robert Gunther-Mohr obituary. *Town Topics,* (Princeton, New Jersey) (March 19, 2014). Accessed November 10, 2016, www.towntopics.com/wordpress/2014/03/19 /obituaries-31914/.

Gilbert, Keith V. *History of the Dayton Project.* Miamisburg, Ohio: Monsanto Research Corporation, 1969. Monsanto Company Records, Washington University Libraries, Department of Special Collections.

Goldberg, Stanley. "General Groves and the Atomic West, the Making and the Meaning of Hanford." In *The Atomic West,* edited by Bruce Hevly and John M. Findlay, 39–89. Seattle: University of Washington Press, 1998.

Groueff, Stephane. *Manhattan Project: The Untold Story of the Making of the Atomic Bomb.* Boston: Little Brown, 1967.

Groves, General Leslie M. *Now It Can Be Told: The Story of the Manhattan Project.* New York: Harper, 1962.

Hewlett, Richard G. and Oscar E. Anderson, Jr. *The New World, 1939/1946: Volume I: A History of the United States Atomic Energy Commission.* University Park: Pennsylvania State University Press, 1962.

Hewlett, Richard G., and Francis Duncan. *Atomic Shield 1947/1952: Volume II of a History of the United States Atomic Energy Commission.* University Park: Pennsylvania State University Press, 1969.

"History of Dayton, Ohio." U-S-History.com. Accessed August 31, 2015, www.u-s-history.com/pages/h2114.html.

Hochwalt, Carroll A. Remarks introducing the medalist of the Industrial Research Institute, "Charles Allen Thomas: The Man and the Scientist" (June 5, 1947). Charles Allen Thomas Papers, Washington University Libraries, Department of Special Collections.

Hoddeson, Lillian, Paul W. Henriksen, Roger A. Meade and Catherine Westfall. *Critical Assembly: A Technical History of Los Alamos During the Oppenheimer Years, 1943–1945.* New York: Cambridge University Press, 1993.

Hammel, Edward F. "The Taming of 49: Big Science in Little Time." *Los Alamos Science,* no. 26 (2000).

Helbert, James E. "Down-to-Earth Explanation Made of Scientific Wonders, Valley Atomic Energy Show Opens with a Bang." *Dayton Journal* (April 28, 1948). Mound Science and Energy Museum archives, Miamisburg, Ohio.

Henry Gabriel Kuivila obituary. *Albany Times Union* (March 18–20, 2004). Accessed October 15, 2016, www.legacy.com/obituaries/timesunion-albany/obituary.aspx?n=henry-gabriel-kuivila&pid=2037751.

"Here is What One Atomic Bomb Would Do If Dropped Near Center of Dayton." *Dayton Daily News* (n.d.).

"High-Flux Years," Oak Ridge National Laboratory *REVIEW,* Numbers Three and Four (1992).

Institute for Atomic Research. Iowa State College. Southwest Museum of Engineering, Communications and Computation website. Accessed November 12, 2016, www.smecc.org/iowa_state_college_-_ames_project_-_laboratory.htm.

"It's Elemental." Jefferson Lab, education.jlab.org/itselemental/ele084.html.

John Walter Schulte obituary. *Albuquerque Journal* (July 11, 2000). Accessed November 6, 2016, obits.abqjournal.com/obits/print_obit/121136.

"Joseph J. Burbage—Personal History." Miami Valley Energy Show Collection. Wright State University Libraries, Special Collections and Archives.

Kany, A. S. "Doomed Runnymede Scene of Many Sports and Cultural Events." *Dayton Herald* (December 30, 1948).

Kathren, Ronald L., Jerry B. Gough, and Gary T. Benefiel, eds. *The Plutonium Story: The Journals of Professor Glenn T. Seaborg, 1939–1946.* Columbus: Battelle Press, 1994.

Keever, Edward. "Pinball Machines Get Atomic Role." *Dayton Herald* (n.d.). Mound Science and Energy Museum archives, Miamisburg, Ohio.

Kelly, Cynthia C. *The Manhattan Project: The Birth of the Atomic Bomb in the Words of its Creators, Eyewitnesses, and Historians.* New York: Black Dog & Leventhal Publishers, 2007.

Kelso, John. "He Helped Put First Atom Bomb Together." *Boston Sunday Post* (March 30, 1947).

"Kentuckian Helps Make Robot Bomb Propellant." *Louisville Courier-Journal* (November 14, 1944).

"Key War Plant Leaders Get Awards." *Dayton Herald* (August 18, 1945).

Koch, Katie. "Faculty Obituaries: A Teacher, Recruiter, and Leader." Boston University *Bostonia* (Summer 2009). Accessed May 15, 2016, www.bu.edu/bostonia/summer09/coulter/coulter.pdf.

Koehnen, Adele U. "Now and Then: Runnymede." *Dayton Daily News* (September 13, 1995).

Lamont, Lansing. *Day of Trinity*. New York: Atheneum, 1965.

Lang, Daniel. "A Reporter at Large: Seven Men on a Problem." *The New Yorker* (August 17, 1946).

Laurence, William L. "Drama of the Atomic Bomb Found Climax in July 16 Test," *New York Times* (September 26, 1945).

Laurence, William L. "Engineering Vision in Atomic Power Project." *New York Times* (October 1, 1945).

"Lawrence and the Bomb." American Institute of Physics, History of Science Web Exhibits. Accessed December 11, 2014, www.aip.org/history/lawrence/bomb.htm.

Lehman, Milton. "The Man Who Made the Bomb." *Esquire* magazine (April 1947): 86.

Los Alamos Scientific Laboratory, Public Relations staff. *Los Alamos, Beginning of An Era, 1943–45*: 34. www.osti.gov/manhattan-project-history/publications/LANLBeginningofEraPart5.pdf.

"Manhattan Project History, Project Y, the Los Alamos Project," Los Alamos Scientific Laboratory, written 1946 and 1947, distributed December 1, 1961. Accessed May 16, 2016, www.osti.gov/manhattan-project-history/publications/LANLMDHProjectYPart1.pdf.

"The Manhattan Project and Predecessor Organizations." American Institute of Physics. Array of Contemporary American Physicists. Accessed April 29, 2016, www.aip.org/history/acap/institutions/manhattan.jsp.

"Memorial Resolution of the Faculty of the University of Wisconsin-Madison on the Death of Professor Emeritus Edwin Merritt Larsen." University of Wisconsin–Madison, Faculty Document 1642 (May 6, 2002). Accessed November 3, 2016, www.secfac.wisc.edu/senate/2002/0506/1642(mem_res).pdf.

"Miamisburg May be Site of New Atomic Research Plant." Newspaper unknown (October 24, 1946). Mound Science and Energy Museum archives, Miamisburg, Ohio.

"Monsanto Abandons Runnymede Lab." Newspaper unknown (December 24, 1948). Mound Science and Energy Museum archives, Miamisburg, Ohio.

"Monsanto Buys Site for 'A' Study." *Dayton Journal* (October 19, 1948).

"Monsanto Chemical Co. Will Produce Robot Launching Propellant." Newspaper unknown (n.d.).

"Monsanto, Duriron Play Important Part in Development of New Bomb." *Dayton Daily News* (August 7, 1945).

"Monsanto Firm's Scientists Aided in Bomb Research." *Knoxville (Tenn.) News-Sentinel* (August 16, 1945).

Monsanto News Release (September 11, 1945). Charles Allen Thomas Papers, Washington University Libraries, Department of Special Collections.

Monsanto News Release (April 24, 1951). Charles Allen Thomas Papers, Washington University Libraries, Department of Special Collections.

"Monsanto Reveals Its Role in Atomic Bomb Research." *Dayton Daily News* (September 14, 1945).

"Monsanto to Push Atomic Research in New Project South of Miamisburg." *Dayton Herald* (October 18, 1946).

Morris L. Nielsen obituary. Ancestry.com. Accessed November 14, 2016, archiver .rootsweb.ancestry.com/th/read/MIKALAMA/2006-10/1159900832.

Moshier, Mary B. "Thomas and Hochwalt Laboratories, Research Division of Monsanto Chemical Company." *Industrial and Engineering Chemistry, Analytical Edition* 10, no. 8 (August 1938): 441–44.

"Mound Ohio, Site." Energy.gov: Office of Legacy Management. Accessed March 24, 2016, www.lm.doe.gov/Mound/Sites.aspx?view=2.

Moyer, Harvey V., ed. "Polonium." Oak Ridge, Tennessee: United States Atomic Energy Commission, Technical Information Service Extension, TID-5221, July 1956.

"Musical Clubs Do Their Best at Spring Concert." *The Tech* (April 21, 1923): 6.

"New Atomic Bomb Draws Explosive Power from Same Source as Sun." *Dayton Daily News* (August 6, 1945, final edition).

"News of Bomb Throws Spotlight on Monsanto and Duriron." *Dayton Daily News* (August 12, 1945): 9.

Norris, Robert S. *Racing for the Bomb: The True Story of Leslie R. Groves, The Man Behind the Birth of the Atomic Age.* New York: Skyhorse Publishing, 2002.

Oak Ridge National Laboratory Review, Numbers Three and Four (1992): 30.

"Ohio University Mourns the Death of Wilfred R. Konneker." Ohio University *Compass* (January 14, 2016). Accessed September 25, 2016, www.ohio.edu/compass/stories /15-16/01/Ohio-University-mourns-the-death-of-Wilfred-R-Konneker.cfm.

Osler, Jack. "Dayton's Wartime Secret Recalled." *Dayton Leisure* (October 13, 1963).

Pagano, Owen N. "The Spy Who Stole the Urchin: George Koval's Infiltration of the Manhattan Project." Undergraduate thesis, The George Washington University, 2014.

"Played Important Role in Releasing Atomic Energy." *Springfield (Mass.) Union-Republic* (October 21, 1945).

"Praise Given Monsanto's Varied War-Aid Program." *Dayton Daily News* (August 18, 1945).

Purchase Agreement: Monsanto Chemical Company/Thomas & Hochwalt Laboratories (February 25, 1936). Monsanto Company Records, Washington University Libraries, Department of Special Collections.

Reed, Bruce Cameron. *The History and Science of the Manhattan Project.* Heidelberg: Springer–Verlag, 2014.

"Resolution of the Corporation of the Massachusetts Institute of Technology on the Death of Charles A. Thomas, Life Member Emeritus" (July 9, 1982). Author's collection.

Roach, Ed. "How Man Handles It—Only the Future Will Know, Dayton and the Manhattan Project." Dayton Aviation Heritage National Historical Park (September 2004).

Rhodes, Richard. *The Making of the Atomic Bomb.* New York: Simon & Schuster, 1986.

"Robert Arthur Staniforth (personal history)." Miami Valley Energy Show Collection. Wright State University Libraries, Special Collections and Archives.

Rock Jr., Lew. "First Atomic Research Plant Photos! Security Outlined at Miamisburg." *Dayton Herald* (n.d.). Mound Science and Energy Museum archives, Miamisburg, Ohio.

Ronald, Bruce W., and Virginia Ronald. *Oakwood: The Far Hills.* Dayton, Ohio: Landfall Press, 1983.

Rosensweet, Alvin. "Runnymede Playhouse Sold to U. S.; Will be Torn Down." *Dayton Daily News* (December 30, 1948).

"Rubber Matters, Solving the World War II Rubber Problem." Chemical Heritage Organization. Accessed May 5, 2016, www.chemheritage.org/research/institute-for-research/oral-history-program/projects/rubber-matters/default.aspx.

"Runnymede is Sold to U.S.; Playhouse will be Destroyed." *Dayton Daily News* (December 30, 1946).

"Runnymede Playhouse is Sold to U.S.; Destruction Planned." *Dayton Daily News* (December 30, 1948).

"Samuel Stimpson Jones." Dunkum Funeral Home. Accessed November 1, 2016, www.dunkumfuneralhome.com/notices/Samuel-Jones.

Sanford, Robert. "Charles A. Thomas: A Voice for Atomic Control." *St. Louis Post-Dispatch* (April 2, 1982).

"Sarasotan Recalls His Work on A-Bomb." *Sarasota (Fla.) Journal* 23, no. 328 (August 18, 1975).

"Science Behind the Atom Bomb." Atomic Heritage Foundation, www.atomicheritage.org/history/science-behind-atom-bomb.

Segrè, Emilio and Clyde Wiegand. "Average Number of Neutrons Emitted per Spontaneous Fission of Pu-240." N.d. [August 10, 1946?] LANL Archive (15–6) 470.1 Plutonium. 2/3/43/–12/28/46. FOIA.

"Sergio De Benedetti—1912–1994." *CMU Physics News*, Inter Actions, 1995/1.

Settle, Frank. "The Manhattan Project." *Analytical Chemistry* (January 1, 2002): 37A.

Shaw, Herbert A. "6700 See Atom Show at Miamisburg." *Dayton Daily News* (n.d.). Mound Science and Energy Museum archives, Miamisburg, Ohio.

Shaw, Herbert A. "Atom Bomb is Energy Incorporated." *Dayton Daily News, Camerica and Magazine* (August 12, 1945).

Shaw, Herbert A. "Compton Sees U.S. Morally Bound to Hold Atom." *Dayton Daily News* (n.d.). Mound Science and Energy Museum archives, Miamisburg, Ohio.

Shaw, Herbert A. "Some Mysteries Will be Unveiled at Miamisburg's World's Fair of Atoms." *Dayton Daily News* (n.d.). Mound Science and Energy Museum archives, Miamisburg, Ohio.

Shook, Howard and Joseph M. Williams. "Building the Bomb in Oakwood." *Dayton Daily News, The Magazine* (September 18, 1983).

"Six Join Faculty for First Year." *Central State Life* (Mount Pleasant, Michigan), 16, 1 (September 26, 1934).

Smith, Alice Kimball. "Behind the Decision to Use the Atomic Bomb: Chicago 1944–1945." *Bulletin of the Atomic Scientists* (October 1958).

Smith, Cyril Stanley. "Plutonium Metal During 1943–45" from *The Metal Plutonium,* edited by A. S. Coffinberry and W. N. Miner. Chicago: University of Chicago Press, 1961.

Smith, Ralph Carlisle. "Comments on Trinity Test Shot Trip" (September 5, 1945). LANL TRINITY Director's Office Files 28. FOIA.

Smyth, Henry DeWolf. *Atomic Energy for Military Purposes: The Official Report on the Development of the Atomic Bomb Under the Auspices of the United States Government, 1940–1945.* Princeton: Princeton University Press, 1945.

Sopka, Katherine R. and Elisabeth M. Sopka. "The Bonebrake Theological Seminary: Top-Secret Manhattan Project Site." *Physics in Perspective* 12, no. 3 (September 2010). Accessed January 15, 2015, link.springer.com/article/10.1007%2Fs00016–010–0019–4#/page-1.

Staley, Samuel R. "Dayton, Ohio: The Rise and Fall and Stagnation of a Former Industrial Juggernaut." *New Geography* (August 4, 2008). Accessed April 1, 2016, www.newgeography.com/content/00153-dayton-ohio-the-rise-fall-and-stagnation-a-former-industrial-juggernaut.

"Staten Island's Tainted Edge, Geologic City Report #3." Accessed March 8, 2016, fopnews.wordpress.com/2010/09/14/staten-islands-tainted-edge-geologic-city-report-3/.

Stockbridge, Dorothy. "Still Going Strong at 80." *St. Louis Globe-Democrat* (August 16–17, 1980): 1F and 3F.

Sublette, Carey. "Section 8.0, The First Nuclear Weapons." Nuclear Weapon Archive. Accessed February 25, 2016, nuclearweaponarchive.org/Nwfaq/Nfaq8.html.

"Summary of the National Defense Research Council." Washington, DC (1946): 507. Accessed March 24, 2016, www.dtic.mil/dtic/tr/fulltext/u2/221610.pdf.

Szasz, Ferenc Morton. *The Day the Sun Rose Twice: The Story of the Trinity Site Nuclear Explosion, July 16, 1945.* Albuquerque: University of New Mexico Press, 1984.

"Text of Truman Secret Weapon Announcement." *St. Louis Post-Dispatch* (August 6, 1945, extra edition).

Thayer, Eliza T. *Katharine Houk Talbott, 1864–1935.* Mount Vernon, New York: Peter Beilenson, 1949.

"Thomas & Hochwalt Laboratories, Chemical Research." Corporate brochure (n.d.). Monsanto Company Records, Washington University Libraries, Department of Special Collections.

Thomas, Charles Allen. "Crossroads." Commencement address (June 13, 1946). Washington University Field House, St. Louis, Missouri. Author's collection.

Thomas, Charles Allen. "The Future is Nearly Here." Speech (June 15, 1961). Rotary Club of Houston. Rice Hotel, Houston, Texas. Charles Allen Thomas Papers, Washington University Libraries, Department of Special Collections.

Thomas, Charles Allen. "The Preparation of Benzene Sulphonyl Chloride and Some of its Derivatives, 1923." Graduate thesis. MIT Archives—Noncirculating Collection 3.

Thomas, Charles Allen. "Priorities in Science." Speech delivered at the 175th Anniversary Convocation at Transylvania College (April 23, 1954). Charles Allen Thomas Papers, Washington University Libraries, Department of Special Collections.

Thomas, Charles Allen. "Radioisotopes—New Tools for Science." Speech (n.d.). Charles Allen Thomas Papers, Washington University Libraries, Department of Special Collections.

Thomas, Charles Allen. "Science and Industry, A Powerful Team." Transcript of address at a dinner honoring Dr. Arthur H. Compton (February 22, 1946). Hotel Jefferson, St. Louis, Missouri. Charles Allen Thomas Papers, Washington University Libraries, Department of Special Collections.

Thomas, Charles Allen. "Science, Progress and the Human Mind." The 26th Steinmetz Memorial Lecture (May 6, 1954). American Institute of Electrical Engineers,

Memorial Chapel, Union College, Schenectady, New York. Charles Allen Thomas Papers, Washington University Libraries, Department of Special Collections.

Thomas, Charles A., and William H. Carmody. "Polymerization of Diolefins with Olefins: I. Isoprene and Pentene-2." *Journal of the American Chemical Society* (June 1932) and *Industrial and Engineering Chemistry:* "Synthetic Resins from Petroleum Hydrocarbons" (October 1932).

Thomas, Charles Allen, and Carroll A. Hochwalt. "Effect of Alkali-Metal Compounds on Combustion," *Industrial and Engineering Chemistry, American Chemical Society* 20 (June 1928): 575.

Thomas, Charles Allen, and John C. Warner. *The Chemistry, Purification and Metallurgy of Polonium.* Oak Ridge, Tennessee: Atomic Energy Commission, Office of Technical Information, 1944.

"Thomas Urges U.S. Keep Dominance in Chemistry." *Dayton Journal* (October 21, 1944).

"Three-Day Atomic Energy Program Opens Today." *Miamisburg News* (April 29, 1948).

Toedtman, J. C. "War Activities of the Central Research Department" (January 20, 1945). Monsanto Company Records, Washington University Libraries, Department of Special Collections.

"Tribute to Dr. Thomas and Mr. Harrington." *St. Louis Post-Dispatch* (December 6, 1963).

"Truman Reveals Mighty Weapon Now Being Used." *Dayton Herald* (August 6, 1945).

"UMD History Connections: Malcolm and the Manhattan Project." University of Maryland Archives (March 10, 2014). Accessed November 12, 2016, umdarchives.wordpress .com/2014/03/10/umd-history-connections-malcolm-the-manhattan-project/.

"University of Illinois: Chemistry, 1941–1951, staff, developments, courses and curricula, publications, doctorate degrees." Accessed November 12, 2016, archive.org /stream/chemistry194119500univ/chemistry194119500univ_djvu.txt.

"U.S. Buys Runnymede Playhouse." *Dayton Herald* (December 30, 1948).

"U.S. Drops New Atomic Bomb on Japs." *Dayton Herald* (August 6, 1945, final edition).

"U.S. Masked Real Purpose for Using Runnymede Site." *Dayton Journal Herald* (December 31, 1948).

"U.S. Synthetic Rubber Program." American Chemical Society. Accessed January 27, 2016, www.acs.org/content/acs/en/education/whatischemistry/landmarks /syntheticrubber.html#us-rubber-research-company.

"U.S. to Raze Runnymede." *Dayton Journal* (December 31, 1948).

U.S. vs. Talbott Realty Company, District Court of the U.S. in and for the Southern District of Ohio, Civil No. 319. (May 23, 1944; September 1946; November 1947).

Vincent, Jack. "Atomic Energy Exhibit Opens at Miamisburg." *Dayton Herald* (April 29, 1948).

Vincent, Jack. "Dr. Hochwalt Asserts U.S. Will Retain Lead in World A-Bomb Race." *Dayton Herald* (April 29, 1948).

"W. C. Fernelius to Tour Southwest in October." *Southwest Retort,* 2, no. 1, American Chemical Society, Dallas/Fort Worth Section (October 1949). University of North Texas Digital Library. Accessed March 1, 2016, digital.library.unt.edu/ark:/67531 /metadc75240/m1/11/.

Waite, George. "Partial Preview Given on Miamisburg Atomic Power Show," *Dayton Herald,* April 19, 1948.

Walsh, Michael. "George Koval: Atomic Spy Unmasked." *Smithsonian* 40, no. 2 (May 2009): 40–47.

Warden, Virginia. "War's Deadliest Bomb Developed on Old Sites of Art and Learning." *Dayton Daily News* (August 12, 1945): 5.

Weimer, Orpha J. "Harry and the Nuclear Threshold." North Manchester Historical Society (NMHS) Newsletter (August 1985 and November 1985). Accessed March 10, 2016, www.nmanchesterhistory.org/biographies-harry-weimer.html.

Wellerstein, Alex. "How Many People Worked on the Manhattan Project?" Restricted Data, The Nuclear Secrecy Blog (November 1, 2013). Accessed April 12, 2016, blog .nuclearsecrecy.com/2013/11/01/many-people-worked-manhattan-project/.

Wellerstein, Alex. "Patenting the Bomb, Nuclear Weapons, Intellectual Property, and Technological Control." History of Science Society: *Isis*, 99 (2008): 57–87.

"What Goes on Here." *Miamisburg News* (March 25, 1948).

"When Dayton Went to War: Memories of the Homefront." PBS, ThinkTV.

Williams, Hill. *Made in Hanford, the Bomb that Changed the World.* Pullman, Washington: Washington State University Press, 2011.

Wright, J. Handly. "Monsanto's New President." *Monsanto* magazine (June 1951). Monsanto Company Records, Washington University Libraries, Department of Special Collections.

Wright, J. Handly, Monsanto News Release (August 7, 1945). Charles Allen Thomas Papers, Washington University Libraries, Department of Special Collections.

Wright, J. Handly, Monsanto News Release (August 14, 1945). Charles Allen Thomas Papers, Washington University Libraries, Department of Special Collections.

"Your Future in the Atomic Age." (n.d.). Monsanto Company Records, Washington University Libraries, Department of Special Collections.

ORAL HISTORIES

Albright, Lyle. Interviewed by author, West Lafayette, Indiana (October 10, 2008).

Bonnet, Vera De Benedetti. Phone interview/email communication by author (December 12, 2015).

Curtis, Mary Lou. Interviewer and date unknown. "Voices of the Manhattan project," Atomic Heritage Foundation. www.atomicheritage.org/profile/mary-lou-curtis.

"Dayton Project—The Initiator.mov," Bill Curtis (May 8, 2012). www.youtube.com /watch?v=J3enrw8LrHY.

De Benedetti, Emma. Phone interview by author, Pittsburgh, Pennsylvania (November 3, 2015).

DuFour, Howard. Interview by James A. Kohler (May 17, 2006). Cold War Aerospace Technology History Project (MS-431). Wright State University Libraries, Special Collections and Archives.

Gates, Ralph. Interviewed by Wendy Steinle, Utah (April 4, 2015). "Voices of the Manhattan project," Atomic Heritage Foundation. www.manhattanprojectvoices.org /oral-histories/ralph-gatess-interview.

Gittler, Max. Interviewed by Alexandra Levy, Florida (December 28, 2012). "Voices of the Manhattan Project," Atomic Heritage Foundation. www.atomicheritage.org /profile/max-gittler.

Gluesenkamp and David Mowry. Interview notes by Frank Curtis (October 1951). "Some Stories from Dayton," Monsanto. Monsanto Company Records, Washington University Libraries, Department of Special Collections.

Hochwalt, Carroll A. Interview notes by Frank Curtis (December 29, 1950). "Background of Thomas & Hochwalt Laboratories: Growth." Monsanto Company Records, Washington University Libraries, Department of Special Collections.

Hochwalt, Carroll A. Transcript of interview by Jeffrey L. Sturchio and Arnold Thackray in Clayton, Missouri (July 12, 1985). Chemical Heritage Foundation: Beckman Center for the History of Chemistry, Oral History Program.

Jones, Betty Halley. Email communication with author (May 27, 2016).

Kistiakowsky, George. Interviewed by Richard Rhodes, Cambridge, Massachusetts (January 15, 1982). "Voices of the Manhattan Project," Atomic Heritage Foundation. www.manhattanprojectvoices.org/oral-histories/george-kistiakowskys-interview.

Leitz, Fred John, Jr. Interviewed by Cameron Foster-Keddie (December 5, 2002). Reed College Oral History Project. Special Collections and Archives, Eric V. Hauser Memorial Library of Reed College.

Mahfouz, George. Interviewed by Edward Roach and Timothy Good of the Dayton Aviation Heritage National Historical Park, Kettering, Ohio (October 7, 2004). Author's collection.

Mahfouz, George. Interviewed by author, Dayton, Ohio (September 8, 2005).

Manhattan Project Veteran Archives. Accessed May 3, 2004, www.childrenofthemanhattan project.org/vet_archives/MPVA_08.htm.

Nichols, Kenneth. Interviewed by Stephane Groueff, Washington, DC (January 4, 1965). "Voices of the Manhattan Project," Atomic Heritage Foundation. www .manhattanprojectvoices.org/oral-histories/general-kenneth-nichols-interview -part-1 and www.manhattanprojectvoices.org/oral-histories/general-kenneth -nichols-interview-part-2.

Setzer, Mr. and Mrs. Interview notes by Frank Curtis (October 1951). "Charles A. Thomas and Thomas & Hochwalt Early Days," Monsanto. Monsanto Company Records, Washington University Libraries, Department of Special Collections.

Sopka, Elisabeth. Phone interview/email communication with author (May 3, 2016).

Thomas, Charles Allen and Mrs. Thomas. Interview notes by Frank Curtis (February 8, 1951). Monsanto Company Records, Washington University Libraries, Department of Special Collections.

Thomas, Charles Allen III. Interviewed by author, La Jolla, California (September 18, 2005).

Weimer, Robert. Phone interview by author, Maryland (March 23, 2016).

Yalman, Richard. Interviewed by Cynthia Kelly, Santa Fe, New Mexico (January 27, 2015). "Voices of the Manhattan Project," Atomic Heritage Foundation. manhattan-projectvoices.org/oral-histories/richard-yalmans-interview.

GOVERNMENT DOCUMENTS

Bainbridge, K. T. "Trinity." Los Alamos Scientific Library (May 1976). LA-6300-H, History. Accessed June 22, 2016, permalink.lanl.gov/object/tr?what=info:lanl-repo /lareport/LA-06300-H.

"Conference on Polonium" at Ryerson Laboratory, University of Chicago. October 20, 1943. LANL (37). FOIA.

Congress of the United Sates, Joint Committee on Atomic Energy. Transcript of Charles Allen Thomas testimony (March 5, 1947). Charles Allen Thomas Papers, Washington University Libraries, Department of Special Collections.

"Directions for Personnel at Campana Hills Camp (Coordinating Council Camp) at Time of Shot" (July 15, 1945). Los Alamos National Laboratory (LANL), TRINITY Director's Office. Files 28. Freedom of Information Act request (FOIA).

"Directions to Personnel at Base Camp at Time of Shot" (July 15, 1945). LANL, TRINITY Director's Office. Files 28. FOIA.

Dodson, Richard W. "Notes on discussions about polonium, Dayton, Ohio—April 11 to 14, 1945." REPT LAMS-239. LANL.

Economides, M., E. Estabrook, and E. F. Joy. "Gamma Scale Chemistry Progress Report." Monsanto Chemical Company—Unit III. U.S. Office of Science and Technical Information Technical Reports (June 1–30, 1948). digital.library.unt.edu/ark:/67531/metadc625194/m1/1/ (Accessed March 8, 2016).

"Health Safety Report: Chemical and Metallurgical Division" (April 1944). REPT LAMS-87. LANL.

"Job and Personnel Summary for the CM Division." (Feb. 1945, and estimates for April 1945). LANL (31–10), 310.1 Laboratory Organization, 4/20/44–9/21/45. FOIA.

Kennedy, Joseph W. "Concerning the First Steps in our Purification of Plutonium" (May 15, 1945). LANL (15–6) 470.1 Plutonium, 2//3/43–12/28/46. FOIA.

Kennedy, Joseph W. "Monthly Progress Report of the Chemistry and Metallurgy Division" (April 1, 1944). Los Alamos Laboratory. REPT LAMS-72-Del. LANL.

Kennedy, J. W. "Monthly Progress Report of the Chemistry and Metallurgy Division" (June 1, 1944). Los Alamos Laboratory. REPT LAMS-97. LANL.

Kennedy, J. W. "Monthly Progress Report of the Chemistry and Metallurgy Division" (November 1, 1944). Los Alamos Laboratory. REPT LAMS-155. LANL.

Kennedy, J. W. "Monthly Progress Report of the Chemistry and Metallurgy Division" (December 1, 1944). Los Alamos Laboratory. REPT LAMS-176-Del. LANL.

Kennedy, J. W. "Monthly Progress Report of the Chemistry and Metallurgy Division" (January 1, 1945). Los Alamos Laboratory. REPT LAMS-190-Del. LANL.

Kennedy, Joseph W. "Notes on chemistry and metallurgy discussions of April 27, 1943." LANL 15–6, 470.1 Plutonium. FOIA.

Kennedy, J. W. and Cyril S. Smith, "Special Report Presented at May meeting with C. A. Thomas on final purification and metallurgy of 49" (May 11, 1944). REPT LAMS-90. LANL.

Kennedy, J. W. and Cyril S. Smith. "Special Report Presented at June meeting with C. A. Thomas on final purification and metallurgy of 49" (June 14, 1944). REPT LAMS-101. LANL.

Kennedy, J. W. and Cyril S. Smith. "Special Report Presented at July meeting with C. A. Thomas on final purification and metallurgy of 49" (July 14, 1944). REPT LAMS-112. LANL.

Kennedy, William R. "Monthly Progress Report of the Chemistry and Metallurgy Division" (October 1, 1944). Los Alamos Laboratory. LAMS-146. LANL.

Kennedy, William R. "Monthly Progress Report of the Chemistry and Metallurgy Division" (December 1, 1944). Los Alamos Laboratory. LANL Research Library. REPT LAMS-176-Del.

"Los Alamos, Beginning of an Era, 1943–45." Los Alamos Scientific Laboratory. Accessed February 1, 2015, www.osti.gov/manhattan-project-history/publications/LANLBeginningofEraPart1.pdf.

"Manhattan District, Scientific Research and Development Personnel (Emilio Segrè)." LANL (20–6) 201.35 Staff Biographies, S-Z ca. 6/45–7/45. FOIA.

"Memorandum of Meeting of July 21, 1943." LANL (37). FOIA.

"Monsanto Chemical Company—Unit III, Progress Report, September 1–15, 1945." National Archives (Atlanta, Georgia). RG 326, 72D2386, Research and Development Division Correspondence Files, 1947–1963, box 14. Folder: Monsanto Chemical Company Report.

"Monsanto Chemical Company—Unit III, Progress Report, 10/15–10/31/45," National Archives (Atlanta, Georgia). RG 326, 72D2386. Research and Development Division Correspondence Files, 1947–1963, box 14. Folder: Monsanto Chemical Company Report.

"Monthly Progress Report of the Chemistry and Metallurgy Division" (May 1, 1944). Los Alamos Laboratory. REPT LAMS-86-Del. LANL.

"Monthly Progress Report of the Chemistry and Metallurgy Division" (February 1, 1945). Los Alamos Laboratory. REPT LAMS-211-Del. LANL.

"Monthly Progress Report of the Chemistry and Metallurgy Division" (March 1, 1945). Los Alamos Laboratory. REPT LAMS-217-Del. LANL.

"Monthly Progress Report of the Chemistry and Metallurgy Division" (May 1, 1945). Los Alamos Laboratory. REPT LAMS-249-Del. LANL.

"Monthly Progress Report of the Chemistry and Metallurgy Division" (June 1, 1945). Los Alamos Laboratory. REPT LAMS-261-Del. LANL.

"Monthly Progress Report of the Chemistry and Metallurgy Division" (July 1, 1945). Los Alamos Laboratory. REPT LAMS-266-Del. LANL.

"Monthly Progress Report of the Chemistry and Metallurgy Division" (August 1, 1945). Los Alamos Laboratory. REPT LAMS-276-Del. LANL.

"Monthly Progress Report of the Chemistry and Metallurgy Division" (October 1, 1945). Los Alamos Laboratory. REPT LAMS-302-Del. LANL.

Mound Laboratory. "Report No. 3 of Steering Committee for Disposal of Units III and IV, Runnymede Road and Dixon Ave. Dayton, Ohio." Atomic Energy Commission (April 17, 1950).

National Institute for Occupational Safety and Health. "It's Elemental." Jefferson Lab, education.jlab.org/itselemental/ele084.html; "Polonium from Irradiated Bismuth Slugs." Evaluation Report Summary: SEC-00049 (May 2006), 5.1.2.

National Park Service, U.S. Department of the Interior. "Manhattan Project Sites Special Resources Study." Manhattan Project Sites, SRS EIS, Newsletter #3 (February 2006).

National Institute for Occupational Safety and Health. "Evaluation Report Summary: SEC-00049, Monsanto Chemical Company" (May 2006).

Ohio Environmental Protection Agency. "PRS 322 Dayton Unit III Soil Screening Results Report" (February 1998).

Oppenheimer, J. Robert. "Minutes of a meeting on polonium, December 16, 1943." LANL (37). FOIA.

Organization Chart, Monsanto Chemical Co., Clinton Laboratories. LANL Archive (10–3) 322 Monsanto 10/5/43–11/4/46. FOIA.

Popham, Richard A. "Health Safety Report: Chemical and Metallurgical Division" (March 1944). REPT LAMS-79. LANL.

Popham, Richard A. "Health Safety Report: Chemical and Metallurgical Division" (May 1944). REPT LAMS-99. LANL.

Popham, Richard A. "Health Safety Report: Chemical and Metallurgical Division" (June 1944). REPT LAMS-108. LANL.

Popham, Richard A. "Health Safety Report: Chemical and Metallurgical Division" (July 1944). REPT LAMS-119. LANL.

Popham, Richard A. "Health Safety Report: Chemical and Metallurgical Division" (August 1944). REPT LAMS-129. LANL.

Popham, Richard A. "Health Safety Report: Chemical and Metallurgical Division" (September 1944). REPT LAMS-143. LANL.

Prestwood, Rene J. "Polonium Extraction Using a Platinum Electrode and Hydrogen," April 10, 1944. LANL (37). FOIA.

Segrè, Emilio. Personnel report (June 13, 1945). "Manhattan District, Scientific Research and Development Personnel (Emilio Segrè)." LANL (20–6) 201.35 Staff Biographies, S-Z ca. 6/45–7/45. FOIA.

"Summary Technical Report of the National Defense Research Committee," Washington, DC (1946). Accessed March 24, 2016, www.dtic.mil/dtic/tr/fulltext/u2/221610.pdf, 478.

Truman, Harry S. "Citation to Accompany the Award of the Medal for Merit to Dr. Charles Allen Thomas" (January 30, 1946). Harry S. Truman Library. Independence, Missouri.

U.S. Atomic Energy Commission. "Retention of Radioactive Property and Salvage Material" (February 10, 1947). AEC Dayton Area, Dayton, Ohio.

U.S. Department of the Air Force, Radiation Safety, Office of Environmental Management with the Ohio Environmental Protection Agency. "Radiological Scoping Survey of Former Monsanto Facilities (Unit III and Warehouse, Dayton, Ohio)." Report date: September 4, 1997, Survey date: August 27, 1997.

U.S. Department of Energy. Gosling, F. G. *The Manhattan Project, Making the Atomic Bomb.* DOE/MA-0001, Washington, DC: History Division, U.S. Department of Energy, January 1999.

U.S. Department of Energy. Gosling, F. G. *The Manhattan Project: Making the Atomic Bomb.* (Revised) DOE/MA-0002 Revised. Washington, DC: U.S. Department of Energy, Office of History and Heritage Resources, National Security History Series, Vol. I. January 2010.

U.S. Department of Energy. "Journal of Glenn T. Seaborg: Vol. 3." Chief, Section C-l, Metallurgical Laboratory, Manhattan Engineer District, 1942–1946; May 1, 1944–April 30, 1945." Glenn T. Seaborg, PUB-112. Lawrence Berkeley Laboratory, University of California. (Reprinted August 1982).

U.S. Department of Energy. *Manhattan District History.* Commissioned in 1944 by General Leslie R. Groves. Prepared by multiple authors under the general editorship of Gavin Hadden of the U. S. Army Corps of Engineers. https://www.osti.gov/opennet/manhattan_district.jsp.

U.S. Department of Energy. "Manhattan District History, Book VIII, Los Alamos Project (Y)—Volume 3, Auxiliary Activities, Chapter 4, Dayton Project" and Appendix, "Historical Report—Dayton Area" (October 31, 1947). D00029699. 20130006684. www.osti.gov/includes/opennet/includes/MED_scans/Book%20VIII%20-%20%20Volume%203%20-%20Auxiliary%20Activities%20-%20Chapter%204,%20Da.pdf.

U.S. Department of Energy, Historic American Engineering Record, Hanford Cultural and Historical Resources Program. "B Reactor (105-B Building)." HAER No. WA-164. Richland, Washington: May 2001. http://wcpeace.org/History/Hanford/HAER_WA-164_B-Reactor.pdf.

U.S. Department of Energy, Office of History and Heritage Resources. "Difficult Choices (1942)," The Manhattan Project, an Interactive History. Accessed March 10, 2016, www.osti.gov/opennet/manhattan-project-history/Events/1942/1942.htm.

U.S. Department of Energy, U.S. Environmental Protection Agency. "Mound Plant, Potential Release Site Package." PRS #320/321/322/323/324/325. OH.45A-4.

U.S. Department of Health and Human Services. "Special Exposure Cohort Petition" (January 9, 2006). Accessed on November 10, 2016, www.cdc.gov/niosh/ocas/pdfs /sec/monsanto/monpet1.pdf.

U.S. Library of Congress, Science Reference Services, Science, Technology & Business Division, Technical Reports and Standards. "Development of the OSRD," The Office of Scientific Research and Development (OSRD) Collection. www.loc .gov/rr/scitech/trs/trsosrd.html.

U.S. Library of Congress, Science Reference Services, Technical Reports and Standards. "The Synthetic Rubber Project." Accessed May 16, 2016, www.loc.gov/rr/scitech /trs/trschemical_rubber.html.

U.S. Department of State. "A Report on the International Control of Atomic Energy" (March 16, 1946). Department of State Publication 2498, Washington, DC.

U.S. Department of State Office of the Historian. "Milestones: 1945–1952, The Acheson-Lilienthal & Baruch Plans, 1946." Accessed March 28, 2016, history.state.gov /milestones/1945–1952/baruch-plans.

War Department, Washington, DC. "Release Prepared as a General News Story in Connection with the Announcement of the Use of the Atomic Bomb Project." LANL TRINITY Director's Office Files 28. FOIA.

War Department, Washington, DC. (Release with background on test of July 16.) LANL TRINITY, Director's Office Files 28. FOIA.

CORRESPONDENCE

Dean Acheson to U.S. Secretary of State. U.S. Department of State, Office of the Historian. Foreign Relations of the United States 1946, Volume I, General; The United Nations (Document 416). 811.2423/3–1746. Accessed April 28, 2016, history.state .gov/historicaldocuments/frus1946v01/d416.

Roger Adams to Charles Allen Thomas. August 23, 1941. Monsanto Company Records, Washington University Libraries, Department of Special Collections.

Roger Adams to Charles Allen Thomas. June 8, 1942. Monsanto Company Records, Washington University Libraries, Department of Special Collections.

Samuel K. Allison to Martin D. Whitaker. July 8, 1943. LANL (37). FOIA.

Robert F. Bacher to J. Robert Oppenheimer. August 14, 1943. LANL (37) FOIA.

Vannevar Bush to Charles Allen Thomas. November 26, 1942. Monsanto Company Records, Washington University Libraries, Department of Special Collections.

T. S. Chapman to J. Robert Oppenheimer. May 21, 1945. LANL (37). FOIA.

Arthur H. Compton to Allison, Oppenheimer, Greenewalt, Seaborg. February 3, 1943. LANL 15–6. 470.1 Plutonium. FOIA.

Arthur H. Compton to J. Robert Oppenheimer. November 29, 1944. LANL (15–7), 470.1, 240Pu, 6/21/44–11/4/46. FOIA.

James B. Conant to S-1 Executive Committee. November 20, 1942. National Archives (College Park, Maryland). Bush-Conant File. Records of the Office of Scientific Research and Development, Record Group 227, 227.3.1 Records of Section S-1 Executive Committee and its predecessors, Microfilm Publication No. M1392—Roll 13—219.

James B. Conant to Charles Allen Thomas. July 28, 1942. Monsanto Company Records, Washington University Libraries, Department of Special Collections.

James B. Conant to Charles Allen Thomas. September 25, 1942. Monsanto Company Records, Washington University Libraries, Department of Special Collections.

James B. Conant to Charles Allen Thomas. July 31, 1943. Monsanto Company Records, Washington University Libraries, Department of Special Collections.

James B. Conant to Charles Allen Thomas. September 24, 1943. National Archives (College Park, Maryland) Bush-Conant File. Records of the Office of Scientific Research and Development, Record Group 227, 227.3.1 Records of Section S-1 Executive Committee and its predecessors, Microfilm Publication No. M1392—Roll 13—219.

James B. Conant to Charles A. Thomas. February 10, 1944. National Archives (College Park, Maryland). Bush-Conant File. Records of the Office of Scientific Research and Development, Record Group 227, 227.3.1 Records of Section S-1 Executive Committee and its predecessors, Microfilm Publication No. M1392—Roll 13—219.

James B. Conant to Charles A. Thomas. July 27, 1944. National Archives (College Park, Maryland) Bush-Conant File. Records of the Office of Scientific Research and Development, Record Group 227, 227.3.1 Records of Section S-1 Executive Committee and its predecessors, Microfilm Publication No. M1392—Roll 13—219.

James B. Conant to Charles A. Thomas. June 4, 1945. National Archives (College Park, Maryland) Bush-Conant File. Records of the Office of Scientific Research and Development, Record Group 227, 227.3.1 Records of Section S-1 Executive Committee and its predecessors, Microfilm Publication No. M1392—Roll 13—219.

James B. Conant to Webster N. Jones, Carnegie Institute of Technology. June 4, 1945. National Archives (College Park, Maryland) Bush-Conant File. Records of the Office of Scientific Research and Development, Record Group 227, 227.3.1 Records of Section S-1 Executive Committee and its predecessors, Microfilm Publication No. M1392—Roll 13—219.

James B. Conant to J. Robert Oppenheimer. July 27, 1944. National Archives (College Park, Maryland) Bush-Conant File. Records of the Office of Scientific Research and Development, Record Group 227, 227.3.1 Records of Section S-1 Executive Committee and its predecessors, Microfilm Publication No. M1392—Roll 13—219.

James B. Conant to Herman B Wells. September 24, 1943. National Archives (College Park, Maryland). Bush-Conant File. Records of the Office of Scientific Research and Development, Record Group 227, 227.3.1 Records of Section S-1 Executive Committee and its predecessors, Microfilm Publication No. M1392—Roll 13—219.

James B. Conant to E. C. Williams, General Aniline and Film Corporation. December 9, 1943. National Archives (College Park, Maryland). Bush-Conant File. Records of the Office of Scientific Research and Development, Record Group 227, 227.3.1 Records of Section S-1 Executive Committee and its predecessors, Microfilm Publication No. M1392—Roll 13—219.

Richard W. Dodson and Emilio Segrè. "On the question of neutrons accompanying alpha decay" (September 22, 1943). LANL 15–6, 470.1 Plutonium. FOIA.

Richard W. Dodson to Robert F. Bacher. May 12, 1945. LANL (37). FOIA and LANL (46–9) 470.1 Polonium, 2/9/44–5/12/45. FOIA.

Richard W. Dodson to Robert F. Bacher. April 24, 1945. LANL (37) and LANL (46–9) 470.1 Polonium, 2/9/44–5/12/45. FOIA.

Richard W. Dodson to C. L. Critchfield. January 12, 1945. LANL (46–9) 470.1 Polonium, 2/9/44–5/12/45. FOIA.

Richard W. Dodson to J. Robert Oppenheimer. August 23, 1944. LANL (37). FOIA.

Richard W. Dodson to J. Robert Oppenheimer. April 30, 1945. LANL (37) and LANL (46–9) 470.1 Polonium, 2/9/44–5/12/45. FOIA.

Richard W. Dodson to Raemer E. Schreiber. May 21, 1945. LANL (37). FOIA.

Dwight D. Eisenhower to Charles A. Thomas. May 5, 1958. Charles Allen Thomas Papers, Washington University Libraries, Department of Special Collections.

W. Conard Fernelius to Norris E. Bradbury. January 3, 1946. LANL (10–3) 322 Monsanto, 10/5/43–11/4/46. FOIA.

W. Conard Fernelius to Norris E. Bradbury, January 24, 1946. LANL (37). FOIA.

W. Conard Fernelius to J. Robert Oppenheimer, August 29, 1945. LANL Archive (10–3) 322 Monsanto, 10/5/43–11/4/46. FOIA.

W. Conard Fernelius to J. Robert Oppenheimer, October 3, 1945. LANL Archive (10–3) 322 Monsanto, 10/5/43–11/4/46. FOIA.

W. Conard Fernelius to J. Robert Oppenheimer, October 11, 1945. LANL Archive (10–3) 322 Monsanto, 10/5/43–11/4/46. FOIA.

J. C. Franklin to Carroll A. Hochwalt. January 19, 1949. Charles Allen Thomas Papers, Washington University Libraries, Department of Special Collections.

Darol Froman to Col. Seeman. July 9, 1946. LANL (37). FOIA.

Leslie R. Groves to Col. E. H. Marsden, District Engineer, Oak Ridge. "Subject: Visit to Site Y." January 30, 1946. LANL (10–3) 322 Monsanto, 10/5/43–11/4/46. FOIA.

Leslie R. Groves to J. Robert Oppenheimer. June 11, 1943. LANL (37). FOIA.

Leslie R. Groves to J. Robert Oppenheimer. August 16, 1943. LANL (37). FOIA.

Leslie R. Groves to J. Robert Oppenheimer. August 15, 1944. LANL (37). FOIA.

Leslie R. Groves to Edgar M. Queeny. August 15, 1945. Charles Allen Thomas Papers, Washington University Libraries, Department of Special Collections.

Leslie R. Groves to Charles A. Thomas. July 29, 1943. LANL Archives (13–9) 400.1144 Historical, 5–10/43–6/19/45. FOIA.

Leslie R. Groves to Charles A. Thomas. August 16, 1943. LANL (37). FOIA.

Leslie R. Groves to Charles A. Thomas. September 29, 1943. LANL (37). FOIA.

Leslie R. Groves to Charles A. Thomas. August 15, 1945. Charles Allen Thomas Papers, Washington University Libraries, Department of Special Collections.

Leslie R. Groves to Charles A. Thomas. November 1, 1945. Charles Allen Thomas Papers, Washington University Libraries, Department of Special Collections.

Joseph G. Hamilton to J. Robert Oppenheimer. May 31, 1943. LANL (37). FOIA.

Joseph G. Hamilton to J. Robert Oppenheimer. July 17, 1943. LANL (37). FOIA.

Robert Ittner to Charles A. Thomas. September 15, 1943. National Archives (College Park, Maryland) Records of the Office of Scientific Research and Development (OSRD), Record Group 227, 1939–50, Bush-Conant File Relating to the Development of the Atomic Bomb, 1940–1945, Section II (DSM—Development of Substitute Materials), Microfilm Publication No. M1392—Roll 13—219.

Eric R. Jette to Norris E. Bradbury. January 29, 1946. LANL (15–6), 470.1 Plutonium 2/3/43–12/28/46. FOIA.

Iral B. Johns to Charles A. Thomas. April 3, 1945. LANL (37). FOIA.

Betty Halley Jones email communication to author. May 27, 2016. Author's collection.

Thomas O. Jones, Corps of Engineers Intelligence Officer to Norris E. Bradbury. January 2, 1946. LANL (10–3) 322 Monsanto, 105/43–11/4/46. FOIA.

Joseph W. Kennedy to Samuel K. Allison. "Request for 95–241 Fraction Processing," June 19, 1945. LANL (13–9) 400.1144 Historical, 5/10/43–6/19/45. FOIA.

Joseph W. Kennedy to J. Robert Oppenheimer. December 29, 1943. LANL (37). FOIA.

George B. Kistiakowsky to Charles A. Thomas. April 17, 1943. Monsanto Company Records. Washington University Libraries. Department of Special Collections.

Wendell M. Latimer to J. Robert Oppenheimer. August 10, 1943. LANL (37). FOIA.

James H. Lum to Richard W. Dodson. May 3, 1944. LANL (37). FOIA.

James H. Lum to Richard W. Dodson. February 5, 1945. LANL (37). FOIA.

James H. Lum to District Engineer, Oak Ridge (E. J. Murphy). October 9, 1945. LANL (10–3) 322 Monsanto, 10/5/43–11/4/46. FOIA.

James H. Lum to J. Robert Oppenheimer. November 9, 1943. LANL (37). FOIA.

James H. Lum to J. Robert Oppenheimer. February 1, 1945. LANL (37). FOIA.

James H. Lum to J. Robert Oppenheimer. February 14, 1945. LANL (37). FOIA.

James H. Lum to J. Robert Oppenheimer. May 11, 1945. LANL (37). FOIA.

James H. Lum to J. Robert Oppenheimer. July 17, 1945. LANL (10–3) 322 Monsanto. FOIA.

James H. Lum to Charles A. Thomas. "Schedule of Postum Deliveries," July 5, 1945. LANL (37). FOIA.

John Manley (Santa Fe) to Charles A. Thomas. October 5, 1943. LANL (10–3) 322 Monsanto. FOIA.

Mary Mead to author. October 17, 2005. Author's collection.

Met Lab to Charles A. Thomas. February 7, 1944. LANL (37). FOIA.

Met Lab (signature page missing) to Charles A. Thomas. January 26, 1944. LANL (37) FOIA.

Ross W. Moshier to James H. Lum, n.d. LANL (37). FOIA.

Kenneth D. Nichols to Charles A. Thomas. May 16, 1945. LANL (37). FOIA.

Kenneth D. Nichols to Charles A. Thomas/James H. Lum. July 5, 1945.

Kenneth D. Nichols to Charles A. Thomas. August 6, 1945. Charles Allen Thomas Papers, Washington University Libraries, Department of Special Collections.

Kenneth D. Nichols to J. Robert Oppenheimer. July 11, 1945. LANL (37). FOIA.

Lothar W. Nordheim to R. L. Doan. "The Case for an Enriched Pile," November 16, 1944. LANL (15–6) 470.1 Plutonium, 2/3/43–12/28/46. FOIA.

J. Robert Oppenheimer to "All Group Leaders Concerned. Subject: Trinity Test." June 14, 1945. LANL TRINITY Director's Office Files 28. FOIA.

J. Robert Oppenheimer to Robert Bacher. August 19, 1944. LANL (37). FOIA.

J. Robert Oppenheimer to Robert Bacher, Richard W. Dodson et al. "Allocation of Plutonium," October 14, 1943. LANL (15–6) 470.1 Plutonium 2/3/43–12/28/46. FOIA.

J. Robert Oppenheimer to T. S. Chapman. June 12, 1945. LANL (37). FOIA.

J. Robert Oppenheimer to Richard W. Dodson. August 29, 1944. LANL (37). FOIA.

J. Robert Oppenheimer to Richard W. Dodson. October 18, 1944. LANL (37). FOIA.

J. Robert Oppenheimer to Leslie R. Groves. May 24, 1943. LANL (9–5) 319.1 Travel Reports, 5/24/43–8/28/45. FOIA.

J. Robert Oppenheimer to Leslie R. Groves. May 27, 1943. LANL 7–4, 319.1 Los Alamos, 5/27/43–4/8/44. FOIA.

J. Robert Oppenheimer to Leslie R. Groves. "Report of the Special Reviewing Committee on the Los Alamos Project," May 27, 1943. LANL (9–4) 319.1 Los Alamos, 5/27/43–4/8/44. FOIA.

J. Robert Oppenheimer to Leslie R. Groves. June 18, 1943. LANL (37). FOIA.

J. Robert Oppenheimer to Leslie R. Groves. July 27, 1943. LANL (37). FOIA.

J. Robert Oppenheimer to Leslie R. Groves. December 18, 1943. LANL (37). FOIA.

J. Robert Oppenheimer to Leslie R. Groves. July 18, 1944. LANL (15–6), 470.1 Plutonium, 2/3/43–12/28/46. FOIA.

J. Robert Oppenheimer to Leslie R. Groves. August 31, 1944. LANL (15–6) 470.1 Plutonium. 2/3/43–12/28/46. FOIA.

J. Robert Oppenheimer to Joseph W. Kennedy and Cyril S. Smith. January 26, 1944.

J. Robert Oppenheimer to L. Langer. May 21, 1945. LANL (37). FOIA.

J. Robert Oppenheimer to Wendell Latimer. July 8, 1943. LANL (37). FOIA.

J. Robert Oppenheimer to Wendell Latimer. August 5, 1943. LANL (37). FOIA.

J. Robert Oppenheimer to James H. Lum. February 9, 1945. LANL (37). FOIA.

J. Robert Oppenheimer to Members of the Coordinating Council. April 20, 1944. LANL (31–10) 310.1 Laboratory Organization, 4/20/44–9/21/45. FOIA.

J. Robert Oppenheimer to Edwin M. McMillan. August 19, 1944. LANL (37). FOIA.

J. Robert Oppenheimer to Bruno Rossi, Robert R. Wilson, et al. September 15, 1945. LANL (10–3) 322 Monsanto, 10/5/43–11/4/46. FOIA.

J. Robert Oppenheimer to Cyril S. Smith. April 23, 1945. LANL (15–6) 470.1 Plutonium, 2/3/43–12/28/46. FOIA.

J. Robert Oppenheimer to Teletype Operators. January 26, 1944. LANL 7–12 322 Monsanto, 10/5/43–11/4/46. FOIA.

J. Robert Oppenheimer to Charles A. Thomas. August 5, 1943. LANL (37). FOIA.

J. Robert Oppenheimer to Charles A. Thomas. September 30, 1943. LANL (37). FOIA.

J. Robert Oppenheimer to Charles A. Thomas. October 5, 1943. LANL (37). FOIA.

J. Robert Oppenheimer to Charles A. Thomas. November 16, 1943. LANL (37). FOIA.

J. Robert Oppenheimer to Charles A. Thomas. February 24, 1944. LANL (37). FOIA.

J. Robert Oppenheimer to Charles A. Thomas. March 14, 1944. LANL (37). FOIA.

J. Robert Oppenheimer to Charles A. Thomas. June 27, 1944. LANL (37). FOIA.

J. Robert Oppenheimer to Charles A. Thomas. September 20, 1944. LANL (37). FOIA.

J. Robert Oppenheimer to Charles A. Thomas. November 3, 1944. LANL (37). FOIA.

J. Robert Oppenheimer to Charles A. Thomas. April 30, 1945. LANL (37). FOIA.

J. Robert Oppenheimer to Charles A. Thomas. June 6, 1945. LANL (37). FOIA.

J. Robert Oppenheimer to Charles A. Thomas. September 8, 1945. LANL (37). FOIA.

Arthur V. Peterson to Norris E. Bradbury. October 1, 1945. LANL (10–3) 322 Monsanto, 10/5/43–11/4/46. FOIA.

F. K. Pittman to David Dow. September 12, 1945. LANL (15–6) 470.1 Plutonium 2/3/43–12/28/46. FOIA.

Rene J. Prestwood to J. Robert Oppenheimer, Joseph W. Kennedy and Emilio Segrè. April 3, 1943. LANL (37). FOIA.

Edgar Queeny to Shareholders of Monsanto Chemical Company. September 8, 1945. Charles Allen Thomas Papers, Washington University Libraries, Department of Special Collections.

John R. Ruhoff, United States Engineer Office, Madison Square Area to U. S. District Engineer Office, Knoxville (Dr. Harry T. Wensel). September 24, 1943. LANL (37). FOIA.

Glenn Seaborg to Charles A. Thomas. October 18, 1968. Monsanto Company Records. Washington University Libraries. Department of Special Collections.

Emilio Segrè to J. Robert Oppenheimer. October 26, 1943. LANL (37). FOIA.

Emilio Segrè to J. Robert Oppenheimer. October 31, 1944. LANL (37). FOIA.

Frank Settle to author. March 1, 2016. Author's collection.

Don Sullenger to Charles A. Thomas III. August 9, 2004. Author's collection.

Edward Teller to Charles A. Thomas. January 4, 1944. LANL (37). FOIA.

Edward Teller to Charles A. Thomas. January 6, 1944. LANL (37). FOIA.

Edward Teller to Charles A. Thomas. June 24, 1944. LANL (37). FOIA.

Edward Teller to Charles A. Thomas. July 28, 1944. LANL (74–15) 319.1 Neutron Emission by Polonium Oxide Layers, 7/28/44.

Charles A. Thomas to Samuel K. Allison. July 30, 1943. LANL (37). FOIA.

Charles A. Thomas to Samuel K. Allison. February 11, 1944. LANL (37). FOIA.

Charles A. Thomas to Captain Benton Bell, Division B, NDRC. April 2, 1943. Monsanto Company Records. Washington University Libraries. Department of Special Collections.

Charles A. Thomas to O. Bezanson. March 13, 1944. Monsanto Company Records. Washington University Libraries. Department of Special Collections.

Charles A. Thomas to Norris E. Bradbury. October 24, 1946. LANL (10–3) 322 Monsanto 10/5/43–11/4/46. FOIA.

Charles A. Thomas to Arthur Compton. October 1, 1943. LANL (37). FOIA.

Charles A. Thomas to James B. Conant. September 21, 1943. National Archives (College Park, Maryland) Microfilm Publication No. M1392—Roll 13–219.

Charles A. Thomas to James B. Conant. January 29, 1944. National Archives (College Park, Maryland) Microfilm Publication No. M1392—Roll 13–219.

Charles A. Thomas to James B. Conant. August 2, 1944. National Archives (College Park, Maryland) Microfilm Publication No. M1392—Roll 13–219.

Charles A. Thomas to James B. Conant. May 18, 1945. National Archives (College Park, Maryland) Microfilm Publication No. M1392—Roll 13–219.

Charles A. Thomas to Enrico Fermi. February 8, 1945. LANL (37). FOIA.

Charles A. Thomas to James Franck, Met Lab. October 1, 1943. LANL (15–6) 470.1 Plutonium, 2/3/43–12/28/46. FOIA.

Charles A. Thomas to James Franck, Met Lab. November 4, 1943. LANL (15–6) 470.1 Plutonium, 2/3/43–12/28/46. FOIA.

Charles A. Thomas to Leslie R. Groves. August 28, 1943. LANL (37). FOIA.

Charles A. Thomas to Leslie R. Groves. September 25, 1943. LANL (37). FOIA.

Charles A. Thomas to Leslie R. Groves. November 3, 1943. LANL (37). FOIA.

Charles A. Thomas to Leslie R. Groves. April 6, 1944. LANL (37). FOIA. National Archives (College Park, Maryland) Microfilm Publication No. M1392—Roll 13–219.

Charles A. Thomas to Leslie R. Groves. July 21, 1944. National Archives (College Park, Maryland) Microfilm Publication No. M1392—Roll 13–219.

Charles A. Thomas to Leslie R. Groves. September 6, 1945. National Archives (College Park, Maryland). Microfilm Publication No. M1392—Roll 13–219.

Charles A. Thomas to Leslie R. Groves. December 17, 1945. LANL (15–8) 470.1 Polonium. 3/30/43–9/18/46. FOIA.

Charles A. Thomas to Leslie R. Groves/James B. Conant. August 23, 1943. LANL Archive (9–4) 319.1 Los Alamos 5/27/43–4/8/44. FOIA.

Charles A. Thomas to Leslie R. Groves/James B. Conant. October 5, 1943. LANL Archive (9–4) 319.1 Los Alamos 5/27/43–4/8/44. FOIA.

Charles A. Thomas to Leslie R. Groves/James B. Conant. "Report No. 3," November 6, 1943. LANL Archive (9–4) 319.1 Los Alamos 5/27/43–4/8/44. FOIA.

Charles A. Thomas to Leslie R. Groves/James B. Conant. "Report No. 4," January 4, 1944. LANL Archive (9–4) 319.1 Los Alamos 5/27/43–4/8/44. FOIA.

Charles A. Thomas to Leslie R. Groves/James B. Conant. "Report No. 5," February 10, 1944. LANL Archive (9–4) 319.1 Los Alamos 5/27/43–4/8/44. FOIA.

Charles A. Thomas to Leslie R. Groves/James B. Conant. "Report No. 6," April 8, 1944. (Covering period February 10, 1944 to April 8, 1944). LANL Archive (9–4) 319.1 Los Alamos 5/27/43–4/8/44. FOIA.

Charles A. Thomas to Leslie R. Groves/James B. Conant. June 13, 1944. LANL (15–6) 470.1 Plutonium 2/3/43–12/28/46. FOIA.

Charles A. Thomas to Major General C. T. Harris Jr. March 26, 1942. Charles Allen Thomas Papers, Washington University Libraries, Department of Special Collections.

Charles A. Thomas to Joseph W. Kennedy. December 1, 1944. LANL (37). FOIA.

Charles A. Thomas to Kenneth D. Nichols. February 8, 1945. LANL (37). FOIA.

Charles A. Thomas to J. Robert Oppenheimer. June 18, 1943. LANL (37). FOIA.

Charles A. Thomas to J. Robert Oppenheimer. July 31, 1943. LANL (37). FOIA.

Charles A. Thomas to J. Robert Oppenheimer. August 5, 1943. LANL (37). FOIA.

Charles A. Thomas to J. Robert Oppenheimer. August 9, 1943. LANL (37). FOIA.

Charles A. Thomas to J. Robert Oppenheimer. September 14, 1943. LANL (15–6) 470.1 Plutonium, 2/3/43–12/28/46. FOIA.

Charles A. Thomas to J. Robert Oppenheimer. February 24, 1944. LANL (37). FOIA.

Charles A. Thomas to J. Robert Oppenheimer. March 22, 1944. LANL (37). FOIA.

Charles A. Thomas to J. Robert Oppenheimer. February 14, 1945. LANL (37). FOIA.

Charles A. Thomas to J. Robert Oppenheimer. March 24, 1945. LANL (37). FOIA.

Charles A. Thomas to J. Robert Oppenheimer. May 14, 1945. LANL (37). FOIA.

Charles A. Thomas to J. Robert Oppenheimer. May 18, 1945. LANL (10–3) 322 Monsanto. FOIA.

Charles A. Thomas to J. Robert Oppenheimer. May 25, 1945. LANL (10–3) 322 Monsanto. FOIA.

Charles A. Thomas to D. M. Sheehan, Monsanto. February 3, 1943. Monsanto Company Records, Washington University Libraries, Department of Special Collections.

Charles A. Thomas to Frances Carrick Thomas. August 15, 1923. Author's collection.

Charles A. Thomas to Frances Carrick Thomas. May 1924. Author's collection.

Charles A. Thomas to Frances Carrick Thomas. August 29, 1945. Author's collection.

Charles A. Thomas to Herman B Wells. September 9, 1943. National Archives (College Park, Maryland) Microfilm Publication No. M1392—Roll 13—219.

Charles A. Thomas III to author. September 18, 2005. Author's collection.

Charles A. Thomas III to author. September 14, 2006. Author's collection.

Charles A. Thomas III to Katharine Bidwell. January 12, 1995. Author's collection.

Charles A. Thomas III to R. Byron Bird. August 9, 1990. Author's collection.

Charles A. Thomas III to Timothy Good. July 23, 2004. Author's collection.

Charles A. Thomas III to Timothy Good. August 1, 2005. Author's collection.

Margaret Thomas to author. December 1, 2005. Author's collection.

Marian Toland to author. October 18, 2005. Author's collection.

John C. Warner to Charles A. Thomas. "Report of a meeting held April 12, 1944 . . . ," April 14, 1944. LANL (15–6) 470.1 Plutonium 2/3/43–12/28/46. FOIA.

Stafford L. Warren to Charles A. Thomas. June 23, 1945. LANL (37). FOIA.

Victor Weisskopf to Lieutenant Taylor. "Eye Witness Account," July 24, 1945. LANL TRINITY Director's Office Files 28. FOIA.

H. T. Wensel to J. Robert Oppenheimer. September 30, 1943. LANL (37). FOIA.

Martin D. Whitaker, Monsanto/Clinton to J. Robert Oppenheimer. September 5, 1945. LANL (10–3) 322 Monsanto, 10/5/43–11/4/46. FOIA.

Martin D. Whitaker, Monsanto/Clinton to A. V. Peterson/Oak Ridge. September 29, 1945. LANL (10–3) 322 Monsanto 10/5/43–11/4/46. FOIA.

E. C. Williams to James B. Conant. December 6, 1943. National Archives (College Park, Maryland) Microfilm Publication No. M1392—Roll 13—219.

E. C. Williams to James B. Conant. December 13, 1943. National Archives (College Park, Maryland) Microfilm Publication No. M1392—Roll 13—219.

TELETYPES

Clear Creek to Dayton. December 18, 1944. LANL (37). FOIA.

Clear Creek to Dayton. December 28, 1944. LANL (37). FOIA.

Clear Creek to Dayton. December 29, 1944. LANL (37). FOIA.

Clear Creek to Dayton. January 5, 1945. LANL (37). FOIA.

Clear Creek to Dayton. May 2, 1945. LANL (37). FOIA.

Clear Creek to Dayton. May 16, 1945. LANL (37). FOIA.

Clear Creek to Monsanto Chem. December 28, 1944. LANL (37). FOIA.

Compton to Oppenheimer. October 9, 1943. LANL (37). FOIA.

Dayton to Clear Creek. February 24, 1944. LANL (37). FOIA.

Dayton to Clear Creek. December 19, 1944. LANL (37). FOIA.

Dayton to Clear Creek. April 30 1945. LANL (37). FOIA.

Dayton to Clear Creek. May 22, 1945. LANL (37). FOIA.

Dayton to Clear Creek. December 28, 1945. LANL (37). FOIA.

Groves (Washington DC) to Clear Creek. August 26, 1944. LANL (37). FOIA.

Groves to Oppenheimer. August 17, 1943. LANL (37). FOIA.

Lum to U.S. Engineer Office, Santa Fe, New Mexico. January 26, 1944. LANL (10–3) 322 Monsanto. FOIA.

Oppenheimer to Conant. July 22, 1943. LANL (37). FOIA.

Oppenheimer to Groves. November 4, 1943. LANL (37). FOIA.

Oppenheimer to Thomas. September 30, 1943. LANL (37). FOIA.

Oppenheimer to Thomas. February 26, 1944. LANL (37). FOIA.

Oppenheimer to Thomas. March 14, 1944. LANL (37). FOIA.

Oppenheimer (Clear Creek) to Area Engineer Chicago (T. S. Chapman). May 28, 1945. LANL (37). FOIA.

Oppenheimer (Clear Creek) to Dayton. August 5, 1944. LANL (37). FOIA.

Oppenheimer (Clear Creek) to Dayton. August 8, 1944. LANL (37). FOIA.

Thomas to Dodson. March 13, 1944. LANL (37). FOIA.

Thomas to Oppenheimer. February 24, 1944. LANL (37). FOIA.

INDEX